Schallum Werner

MicroRNA processing in Arabidopsis thaliana

Schallum Werner

MicroRNA processing in Arabidopsis thaliana

An extensive in vivo structure-function analysis of A. thaliana pri-miRNA 172a

Südwestdeutscher Verlag für Hochschulschriften

Imprint

Any brand names and product names mentioned in this book are subject to trademark, brand or patent protection and are trademarks or registered trademarks of their respective holders. The use of brand names, product names, common names, trade names, product descriptions etc. even without a particular marking in this work is in no way to be construed to mean that such names may be regarded as unrestricted in respect of trademark and brand protection legislation and could thus be used by anyone.

Publisher:
Südwestdeutscher Verlag für Hochschulschriften
is a trademark of
Dodo Books Indian Ocean Ltd., member of the OmniScriptum S.R.L Publishing group
str. A.Russo 15, of. 61, Chisinau-2068, Republic of Moldova Europe
Printed at: see last page
ISBN: 978-3-8381-2018-8

Zugl. / Approved by: Tübingen, Eberhard Karls Universität, Diss., 2010

Copyright © Schallum Werner
Copyright © 2010 Dodo Books Indian Ocean Ltd., member of the OmniScriptum S.R.L Publishing group

ACKNOWLEDGEMENTS

First of all I would like to thank and appreciate the help and guidance of my advisor Detlef Weigel. Without his support and encouragement this PhD thesis would not have been possible! I would also like to thank him for giving me the opportunity to go to the US and work in Nina Fedoroffs laboratory. I will never forget that exceptional experience.
I also want to thank my PhD committee Silke Hauf and Friedrich Schöffl for advice and help during my PhD.

Further I would like to thank Sascha Laubinger, Marco Todesco and Heike Wollmann to help me finding a new path when all the old ones seemed to have ended.
I thank Nina Fedoroff especially for giving me the chance to stay and work 6 weeks in her lab. And Liang Song for sharing precious data, watching Star Trek 11 and a great time in State College.
Thanks to the whole miRNA gang for being a social spotlight in the scientific world. And of course for the cake, the fun, the parties, the gossip, …

Thanks to Heike for being a great and fantastic bench-neighbour (I am sure it was not always easy ☺), to Felipe for the important discussions about science, life, the universe and everything, to Sascha for answering all my silly questions and to be my "???"-dealer, to Patrice not only for finding for me the most important book in life but helping me in several situations, to Eva, Stephanie and Verena for several good laughs and movie nights, to Lisa for helping me with English grammar, to Hülya for being so patient with me, to Joe for a lot of off-colour jokes, to Frank the best Lab-Manager ever ;-) , to Levi and Kirsten for advise on how to buy a good bike and the first bike ride in the Schönbuch, to Pablo for joining our lab and being himself, to Nacho, Roosa, Eunyoung, Sang-Tae, Beth, Vini, Janina, Anette, Helena, Anna-Lena, Markus, Wolfgang, Dani, Yasushi, Christina, Carolin, Linda, Regina, Gabi and Korbinian for … countless things. And Julia for the Lago Maggiore!

Thanks to Nadine Wittkopp for having a great time together when I was writing this thesis, and giving me coffee from Dep. II coffee machine.

VERY !!!!! special thanks to Rebecca Schwab, Sascha Laubinger, Heike Wollmann and Aurelia Fuchs, the audiobook club!!! Thanks also to Justus, Peter and Bob.... you know why!

Thanks to Claudia for the beginning.
And thanks to my family for ... everything.

P.S.: Of course it was because of the ice-machine!!!

PUBLICATIONS

During the course of this work, the following article has been published:

Werner, S., Wollmann, H., Schneeberger, K. and Weigel, D. (2010). Structure Determinants for Accurate Processing of miR172a in *Arabidopsis thaliana*. Current Biology **20**, 42-48.

CONTRIBUTIONS

Heike Wollmann did the initial EMS screen on miR172a overexpressors, from which 3 mutants were isolated (HW120, HW121, HW122). Korbinian Schneeberger did the systematic survey of conserved *A. thaliana* miRNAs.

Table of Contents

Acknowledgements 1
Publications 2
Contributions 2
Summary 4
General Introduction 5
The Beginning 5
Several classes of small RNAs 6
Small RNA biogenesis 7
A closer look to microRNAs 9
Chapter I 15
Introduction 15
Results 16
Discussion 19
Chapter II 21
Introduction 21
Results 23
A common sequence binding motif in the miRNA precursors? 23
A landmark in the energy profile of miRNA foldbacks? 24
Discussion 28
Chapter III 29
Introduction 29
Results 30
Effects of point mutations on pri-miR172a processing efficiency 30
Processing determinants in the proximal region of miR172a foldback 31
Processing determinants in the distal region of miR172a foldback 34
Design of a minimal miRNA 34
Discussion 39
References 41
Materials and Methods 51
Supplementary Material 56
Inventory of Supplemental Information 56

SUMMARY

During the last decade small RNAs came more and more into focus in developmental biology. They turned out to play not only a role in defence mechanisms but also in guiding and restricting developmental processes of eukaryotic organisms, both animals and plants.

A particularly important class of small RNAs are microRNAs. Plant microRNAs (miRNAs) have high sequence complementarity to their targets, and they are thought to regulate target mRNAs mainly by cleavage, in contrast to animal miRNAs, which mainly inhibit translation.

Plant miRNAs are processed from a longer self-complementary precursor by the RNase III-like enzyme DICER-LIKE1 acting in concert with the double-stranded RNA-binding protein HYPONASTIC LEAVES1 and the zinc finger protein SERRATE. Together, they excise a miRNA duplex with a characteristic 3' two-nucleotide overhang from the primary miRNA transcript (pri-miRNA). In animals pri-miRNAs are structurally very homogenous, with a stereotypic position of the miRNA within a foldback. Accordingly, rules for miRNA excision from the precursor are quite simple in animals. In contrast, how miRNA sequences are recognised in the structurally much more diverse foldbacks of plants has been previously unknown. I have performed an extensive in vivo structure-function analysis of *Arabidopsis thaliana* pri-miRNA172a (pri-miR172a). A junction of single-stranded to double-stranded RNA 15 nucleotides proximal from the miRNA duplex appears to be essential for accurate miR172a processing. This attribute is found in several other but not all plant miRNA foldbacks. In addition, I have identified structural features of the distal foldback important for miR172a processing. Our ability to engineer *de novo* a functional minimal miRNA precursor highlights that I have discovered several elements both necessary and sufficient for accurate miRNA processing.

In addition I found indications that the stability of miRNA foldbacks likely plays a role in miRNA processing.

GENERAL INTRODUCTION

The Beginning

In 1993 Lee et al. and Wightman et al. could show that in the nematode *Caenorhabditis elegans* (*C. elegans*) the DNA locus *lin-4* acts as a negative regulator of the heterochronic gene *lin-14* by producing small RNAs (Lee et al., 1993; Wightman et al., 1993). These small RNAs are approximately 22 nucleotides (nt) long and are partially complementary to a repeated sequence element in the 3' untranslated region (UTR) of *lin-14*. In this work, it was shown for the first time, that endogenous small non-coding RNAs could possibly interact with a messenger RNA (mRNA) and therefore influence the levels of the protein produced from the mRNA posttranscriptionally. Unfortunately, the impact of this discovery was largely ignored for several years. Later on, the regulatory mechanism was described in more detail, confirming that indeed *lin-4* small RNAs (sRNAs) form a duplex with elements in the *lin-14* 3' UTR (Ha et al., 1996) and that *lin-4* controls more than one gene (Moss et al., 1997). However, it was thought that this is just a peculiar and exceptional mode of RNA-mediated regulation specific to nematodes.

Fire and Mello reported in 1998 a technique to silence endogenous genes in *C. elegans* (Fire et al., 1998). It was known that in some biological systems injected long single-stranded sense or antisense RNA can influence the function of a corresponding gene (Fire et al., 1991; Izant and Weintraub, 1984; Nellen and Lichtenstein, 1993). It was suggested, that this is mediated via a simple complementary hybridisation mechanism between the exogenous RNA and endogenous mRNA. Therefore, the mechanism was called RNA interference (RNAi). Fire and Mello also tried to interfere with gene function by injecting long (300 – 1.000 bases) complementary RNA into *C. elegans*. To their surprise, double-stranded RNA (dsRNA) was much more effective than sense or antisense single-stranded RNA (ssRNA). Additionally, the degree of silencing was not a stoichiometric effect but more like triggering a cascade. For their discovery of this silencing phenomenon they were awarded with the Nobel Prize in 2006.

In 1999 Hamilton and Baulcombe showed in plants, that a seemingly unrelated silencing phenomenon, called PTGS (posttranscriptional gene silencing), correlated with the presence of 25 nt small RNAs derived from both the sense and antisense strand. PTGS targets and silence transgenic and viral mRNA (Cogoni et al., 1996; Matzke et al., 1989;

Napoli et al., 1990). The observed small antisense RNAs were complementary to the silenced mRNA (Hamilton and Baulcombe, 1999). They were only detected together with transgene sense transcription or virus-replication. Slowly, it seemed that small RNAs play a much bigger role in transcript regulation than initially thought.

In the year 2000 two other groups showed in *Drosophila in vitro* system or cell-culture, that exogenous long dsRNA was processed into sRNAs (21-25 nt) during RNAi (Hammond et al., 2000; Zamore et al., 2000). Cell fractions containing these sRNAs had enzymatic silencing activity and these now called short interfering RNAs (siRNAs) were responsible for the specificity of gene silencing (Elbashir et al., 2001).

During the same time, a second endogenous sRNA (*let-7*) was described in *C. elegans*, which also negatively regulates its target mRNA *lin-41* (Reinhart et al., 2000). Shortly afterwards, it was clear that *let-7* is conserved among several species, including vertebrate, ascidian, hemichordate, mollusc, annelid and arthropod (Pasquinelli et al., 2000). With these data, the question came up, if the new post-transcriptional silencing mechanism was really an exception? Scientists started to pay more attention to the bottom of their polyacrylamide gels and began to clone and sequence small RNAs. In an exceptional issue of Science three groups could proof, that there are much more endogenous small RNAs (now called microRNAs) than initially thought. They cloned several more conserved microRNAs, not only from *C. elegans*, but also from other invertebrates and vertebrates (Lagos-Quintana et al., 2001; Lau et al., 2001; Lee and Ambros, 2001). A door to a new world of small regulatory RNAs was pushed open.

Several classes of small RNAs

During the last years several sRNA classes were described in plants. Especially deep-sequencing approaches revealed for the first time the complexity of sRNAs classes (Fahlgren et al., 2007; Rajagopalan et al., 2006). Next to miRNAs and siRNAs also trans acting short interfering RNAs (ta-siRNAs), natural antisense short interfering RNAs (nat-siRNAs) and repeat associated short interfering RNAs (ra-siRNAs) do exist.

Ta-siRNAs were first described in company with new alleles of the genes RNA-DEPENDENT-RNA-POLYMERASE6 (RDR6) and SUPPRESSOR OF GENE SILENSING3 (SGS3), which were isolated in EMS, fast-neutron and T-DNA screens (Peragine et al., 2004). One gene was specifically upregulated in this mutants and it was shown that it is post-transcriptionally downregulated by trans-acting siRNAs. In a parallel publication ta-siRNAs were described more precisely, together with proteins involved in ta-siRNA biogenesis, which are not only RDR6 and SGS3 but also components of the miRNA pathway (Vazquez et al., 2004b). Indeed, to generate ta-siRNAs a miRNA has to cleave initially a non-coding mRNA (Allen et al., 2005). RDR6 and SGS3 form and stabilize a double-stranded RNA out of these fragments and the RNase III-like enzyme DICER-LIKE4 processes the double-strand into 21 nt ta-siRNAs (Howell et al., 2007). RNase III-like enzymes and sRNA biogenesis is described in more detail below.

24 nt long nat-siRNAs, as the name indicates, derive from genes, which natural *cis* anti-sense transcripts partially overlap (Borsani et al., 2005). The first described nat-siRNAs come from a gene involved in stress tolerance and a gene with unknown function. Their anti-sense transcripts form a dsRNA and give rise to 24 nt nat-siRNAs. These nat-siRNAs target the stress-gene again and trigger phased production of 21 nt nat-siRNAs.

Ra-siRNAs are also called heterochromatic siRNAs. They were the first time described in a publication using a genetic approach to identify three distinct sRNA-generating pathways (Xie et al., 2004). The DCL3/RDR2 pathway generates ra-siRNAs, which target transposons, retroelements or trigger DNA methylation.

However, the focus of this PhD-thesis is on miRNAs and after a general introduction into sRNA biogenesis miRNAs are described into more detail.

Small RNA biogenesis

The sources of sRNAs in *Arabidopsis thaliana* (called here Arabidopsis) are very variable. They can derive from virus replication, transcription of inverted repeats, convergent transcription or directly from non-coding endogenous genes. Whatever the source is, the derived product is a partially or full complementary double-stranded RNA. An RNase III-like nuclease enzyme of the Dicer family processes sRNA duplexes out of these dsRNA (Bernstein et al., 2001) in concert with co-effector proteins. Arabidopsis has 4 members of the Dicer family, DICER-LIKE1 – 4 (DCL1 - 4) (Schauer et al., 2002). Typically DCL proteins contain several domains (in this order): DExH-helicase, helicase C, DUF283, PAZ, two RNase III (a and b) and two double-stranded RNA-binding domains (dsRBD) (Figure 01) (Margis et al., 2006). The length of the resulting sRNA duplex (18 – 25 nt) depends on the physical distance between the PAZ domain and the two RNase III domains.

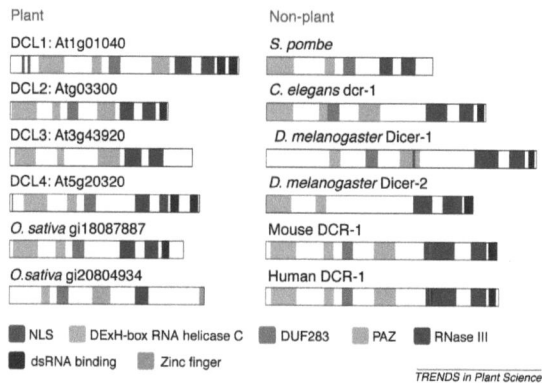

Figure 01: Dicer family members in different organisms (Schauer et al., 2002)
Protein domains are indicated by colours.

The PAZ domain can specifically bind two nucleotide 3' overhangs of dsRNA ends. The substrate RNA extends approximately two helical turns along the protein surface before it reaches a single processing center. The two RNase III domains "a" and "b" form an intramolecular dimer where each domain cuts one of the RNA strands leaving a 5' monophosphate at the product ends and (again) a two nt 3' overhang (MacRae et al., 2007; Macrae et al., 2006). In plants, the processed RNA duplex is afterwards 2'-O-methylated by the S-adenosyl methionine-dependent methyltransferase HUA ENHANCER 1 (HEN1) (Huang et al., 2009; Li et al., 2005; Yang et al., 2006b), protecting the 3' end from uridylation and 3'-to-5' exonuclease mediated degradation (Li et al., 2005; Ramachandran and Chen, 2008). Subsequently, the RNA duplex is exported from the nucleus into the cytoplasm via the Exportin5 ortholog HASTY and other unknown factors (Park et al., 2005).

In the cytoplasm the guiding strand of the sRNA duplex is loaded into one of several RNA-induced silencing complexes (RISCs). All of them contain a member of the ARGONAUTE (AGO) protein family as catalytic subunit. AGO proteins are characterised by a PAZ (binding RNA 3' end), Piwi (catalytic activity) and Mid domain (binding selectively 5' nucleotides with monophosphates). In Arabidopsis 10 AGO proteins have been identified (Vaucheret, 2008) with AGO1 incorporating miRNAs and siRNAs to repress and/or cleave its targets (Baumberger and Baulcombe, 2005; Brodersen et al., 2008; Qi et al., 2005), AGO4 and AGO6 functioning in sRNA mediated transcriptional gene silencing (TGS), and

AGO7 having a role in ta-siRNA function (Montgomery et al., 2008). Further, some of the AGO proteins show a binding preference for certain 5' end nucleotides, for example, AGO1 for uridine (Mi et al., 2008; Montgomery et al., 2008).

After the guide sRNA is loaded into the RISC, it interacts with complementary nucleic acids to execute its function: a) endonucleolytic cleavage of sRNA-target hybrids, b) translational repression through unclear mechanisms (Brodersen et al., 2008) and c) guiding DNA cytosine and/or histone methylation (Herr et al., 2005).

A closer look to microRNAs

As mentioned above, several small RNA classes were described in plants: siRNAs, ta-siRNAs, nat-siRNAs or ra-siRNAs (Jamalkandi and Masoudi-Nejad, 2009; Voinnet, 2009).

Among these small RNAs, microRNAs (miRNAs) have received particularly intense attention, as several miRNA biogenesis mutants show severe defects or are embryonic-lethal (Schauer et al., 2002; Vaucheret et al., 2004). MiRNAs can originate from introns of protein coding mRNAs or polycistronic RNAs from intergenic regions. However most of them are driven by their own promoter and transcribed by the RNA Polymerase II (Figure 02) (Lee et al., 2004; Xie et al., 2005). Therefore the primary microRNA transcript (pri-miRNA) contains a 7-methyl guanosine 5' CAP structure and a polyadenosine 3' end (polyA). In Arabidopsis *dawdle* (*ddl*) mutants the levels of several pri-miRNAs and their corresponding mature miRNAs are reduced (Yu et al., 2008). It has been proposed that the DAWDLE protein stabilizes pri-miRNAs or guides the processing protein DCL1 to certain pri-miRNA transcripts due to the direct interaction between DDL and DCL1. The processing of *Arabidopsis* miRNAs occurs in so-called nuclear "dicing-bodies" (D-bodies) or SmD3/SmB-bodies (Fang and Spector, 2007; Fujioka et al., 2007; Song et al., 2007).

General Introduction

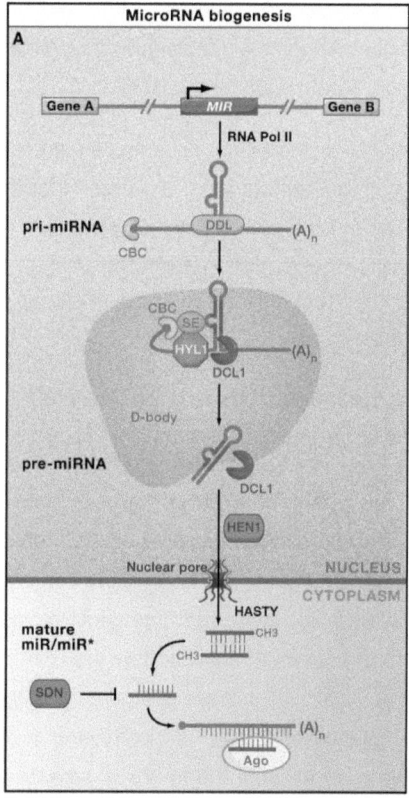

Figure 02: MicroRNA biogenesis in Arabidopsis (Voinnet, 2009)
MiRNA genes are transcribed by RNA Pol II. The transcript (pri-miRNA) forms an internal foldback structure, which is processed in D-bodies by the DCL1-processing complex. The excised miRNA duplex is 3' methylated and transported into the cytoplasm. The guide strand is incorporated into the AGO-silencing complex, which cleaves the complementary target mRNA.

The pri-miRNA forms a partially complementary double-stranded foldback structure, with a diverse length in plants. Similarly the secondary structure varies. This foldback structure (Figure 03) is recognised by the RNase III-like nuclease DCL1, the double-stranded RNA-binding protein HYL1 and the C2H2-zinc finger protein SE (Grigg et al., 2005; Han et al., 2004; Kurihara and Watanabe, 2004; Lobbes et al., 2006; Prigge and Wagner, 2001; Vazquez et al., 2004a; Yang et al., 2006a) leading to the processing of the RNA molecule. Several *in vitro* reports demonstrated that DCL1 alone is able to generate small RNAs, but accurate processing of miRNAs is only possible in concert with HYL1 and SE (Dong et al.,

2008; Kurihara et al., 2006). The exact role of HYL1 and SE is still not entirely clear. It might be that HYL1 binds pri-miRNAs and positions the catalytic enzyme DCL1 precisely on the miRNA precursor. Interaction studies between DCL1 and HYL1 support this model (Kurihara et al., 2006). SE on the other hand plays not only a role in miRNA processing, but also in mRNA splicing, since weak *se* mutants show similar mRNA splicing defects as the mutated cap-binding proteins (CBP) *abh1/cbp80* and *cbp20* (Laubinger et al., 2008), and *abh1/cbp80* and *cbp20* show similar defects and phenotypical characteristics compared to *se* mutants. Indeed it has been shown that also the cap-binding complex (CBC) plays a role in miRNA processing: several but not all miRNA levels are reduced and the corresponding precursors are upregulated in *abh1/cbp80* and *cbp20* mutants (Gregory et al., 2008; Kim et al., 2008; Laubinger et al., 2008). However, the requirement of *HYL1* and *SE* in miRNA biogenesis seems to be not the same for all the plant miRNAs. Strong mutant alleles of these two genes are still able to produce normal levels of several miRNAs.

After the miRNA duplex is released the 3' ends are methylated by HEN1 to avoid degradation. The duplex is transported into the cytoplasm and the miRNA (guide strand) is loaded into the AGO1 RNA-induced silencing complex (see above). There are very few examples where miRNAs do not bind to AGO1 but other AGO proteins, like miR390 binding to AGO7 (Montgomery et al., 2008). As mentioned above, different AGO proteins prefer small RNAs with specific 5' nucleotides: AGO1 prefers uridine, AGO2 and AGO4 to adenosine and AGO5 to cytosine (Mi et al., 2008; Montgomery et al., 2008; Takeda et al., 2008).

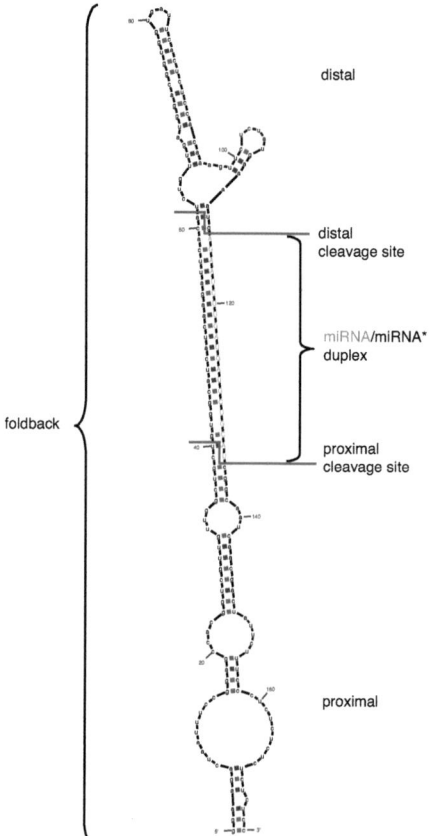

Figure 03: Secondary structure of miR172a foldback
The miRNA is indicated in green, DCL1 cleavage sites are marked in blue.

Changing the 5' terminal nucleotide can guide a miRNA into a different AGO complex. After the guide strand is loaded into the RISC the passenger strand (miRNA*) is in most cases degraded fast.

In plants miRNAs guide the RISC to the highly complementary target mRNA and trigger the slicing function of AGO1 proteins. The remaining mRNA fragments are degraded by exoribonucleases (Baumberger and Baulcombe, 2005; Gy et al., 2007; Souret et al., 2004). Although most of the miRNAs cause cleavage of their targets, translational repression was one of the first reported effects (Aukerman and Sakai, 2003; Brodersen et al., 2008; Chen, 2004; Gandikota et al., 2007). At the moment, the scientific community discusses how often translational repression occurs in plants and if it is an occasional event of a few miRNAs or a general mechanism acting in combination with target cleavage (Voinnet, 2009).

MiRNAs influence nearly every developmental process or regulatory mechanism in plants. For example, miRNA156 and miR172 are involved in regulating flowering time (Aukerman and Sakai, 2003; Wang et al., 2009; Wu et al., 2009), miR319 in hormone biosynthesis (Schommer et al., 2008), miR160 in seed germination (Liu et al., 2007), miR164 in shoot development (Sieber et al., 2007), miR393 in bacterial defence (Ruiz-Ferrer and Voinnet, 2009) and miR172 also in regulation of shoot apical floral stem cells (Zhao et al., 2007). Although the understanding of miRNA biogenesis has increased during the last years, we still know little about its regulation. It has been shown that the two main proteins in the miRNA pathway themselves (DCL1 and AGO1) are targets of miRNAs (Vaucheret et al., 2004; Xie et al., 2003). Mature miRNAs on the other hand are under a tight control by exonucleases (Ramachandran and Chen, 2008). Additionally, it has been observed that short interspersed repetitive elements (SINE) can function as negative regulators of biogenesis machinery. They form foldback structures mimicking a pri-miRNA secondary structure. HYL1 can bind to these elements and is therefore competed out of the miRNA biogenesis pathway (Pouch-Pelissier et al., 2008).

At the beginning of this PhD the protein components of the miRNA biogenesis pathway had already been identified. However, how the DCL1-processing machinery recognises the correct mature miRNA nucleotides in the pri-miRNA transcript was still unknown. It was the aim of my thesis to answer that question.

A first idea came from a paper describing how the processing complex is positioned on the miRNA foldback in *Drosophila melanogaster* (Han et al., 2006). Here a typical pri-miRNA foldback is approximately 33 bp long. A single-stranded to double-stranded RNA junction with a following 11 bp full complementary double-stranded stem positions the processing complex at the correct proximal processing site. This pri-miRNA to pre-miRNA processing step results in a stem-loop with a defined proximal miRNA duplex end (5' and 3' of the stem loop) and a still attached loop structure at the distal end of the duplex. In animals the second processing step happens in the cytoplasm (reviewed in Winter et al., 2009), meaning that the pre-miRNA is exported from the nucleus and cut again 21 nt distal from the first proximal processing site, releasing a mature miRNA duplex.

In plants, miRNA processing occurs only in the nucleus, and there is no evidence for a similar spatially segregated two-step mechanism (reviewed in Voinnet, 2009). Additionally plant miRNA foldbacks are much more diverse in structure and length than animal foldbacks (Xie et al., 2005). Here, I refer to the miRNA containing folded structure in the pri-miRNA as foldback (Figure 03; Figure S1a, S1b). To address the question how the DCL1-complex recognizes the correct miRNA sequence, I generated a reporter construct to monitor processing efficiency of mutated pri-miRNAs (see Chapter I). Additionally I analysed the sequence and the secondary structure of several pri-miRNAs, including the energy-profile of the RNA structures (see Chapter II). I also introduced point mutations into the RNA foldback of miR172a (our model system), to pinpoint processing-sensitive regions in the foldback (see Chapter III).

CHAPTER I

INTRODUCTION

To observe how miRNAs are generated, it would be useful to have a tool that is able to monitor the efficiency of pri-miRNA processing not only for one, but several miRNAs. This tool would make it possible to compare processing efficiency of endogenous or artificial pri-miRNA foldbacks and to analyse, if very well processed foldbacks share common structural characteristics. Additionally, mutant screens, which would check for components involved in miRNA biogenesis, could be performed much faster and easier. Furthermore, it would be interesting to test processing variation within different tissues. A specific miRNA might be expressed equally in several tissues but processed differently and therefore lead to a variation in miRNA activity. The targets of this miRNA would be heavily repressed in tissues with high miRNA processing efficiency but only moderately repressed in tissues with low processing efficiency/activity. A standard miRNA promoter-GFP fusion construct cannot be used to monitor this processing variation. Such reporter would indicate transcription differences, but no fluctuation or changes in processing events, because it does not enter the miRNA-processing pathway. I therefore wanted to develop a new reporter construct, which could monitor such differences in miRNA processing.

I designed two constructs overexpressing triple GFP or Luciferase, respectively. The 3' UTR of these constructs encoded for the miRNA precursor pri-miR172a (Figure 04). Overexpression of miR172a leads to very early flowering plants (2 leaves). This phenotype is convenient to analyse and can be quantified using the number of rosette leaves as a proxy. I expected that in wild-type plants, the transcript would be cleaved due to processing of the pri-miRNA positioned in the 3' UTR. The remaining fragments would be quickly degraded by exonucleases (Gy et al., 2007). Thus, the reporter transcript would not be translated and no signal could be detected (Figure 04). In contast, in a miRNA biogenesis mutant background, I expected that processing of the miRNA precursor would be abolished, the plant would flower normally (10-12 leaves) and the reporter transcript could be translated and its signal detected.

RESULTS

I tested the approach described above with the 35S::Luc--pri-miR172a construct. As a positive control I used the *Drosophila* pri-miRNA bantam as 3' UTR, as it is not processed in Arabidopsis (Figure 05) and the fluorescence derived from the reporter should be easily detectable.

As expected the untransformed wild-type Col-0 plants had just background luciferase activity compared to the positive control. Unfortunately, the tested 35S::Luc--pri-miR172a construct had a three times higher luciferase activity than the positive control (and 2,000 times higher than the untransformed Col-0). Beyond that, the variation was extremely high in all cases. Based on these facts, I decided to not use the luciferase system to monitor pri-miRNA processing efficiency.

I further tested the corresponding GFP constructs in plants (35S::GFP--pri-miR172a) (Figure 06). As expected, GFP was not detected in wild-type Col-0 plants because of the processed miRNA precursor in the 3' UTR. In miRNA biogenesis mutants on the other hand, GFP could be detected: the mutants *dcl1* and *se* accumulated GFP. Unexpectedly, *hyl1* mutant plants did not express GFP, although HYL1 plays an important role in miRNA biogenesis (Han et al., 2004). In *hyl1* mutants mature miRNA levels are reduced and the precursors are upregulated (Han et al., 2004).

As a control, I examined *dcl2,3,4* mutants, which did express GFP. The Dicer-like proteins DCL2, DCL3 and DCL4 do not play a role in miRNA processing and hence, the miR172a should have been released from the construct, preventing GFP accumulation. Obviously, this was not the case as indicated by a strong GFP signal.

This project was finally terminated, due to the puzzling results, which I was not able to interpret.

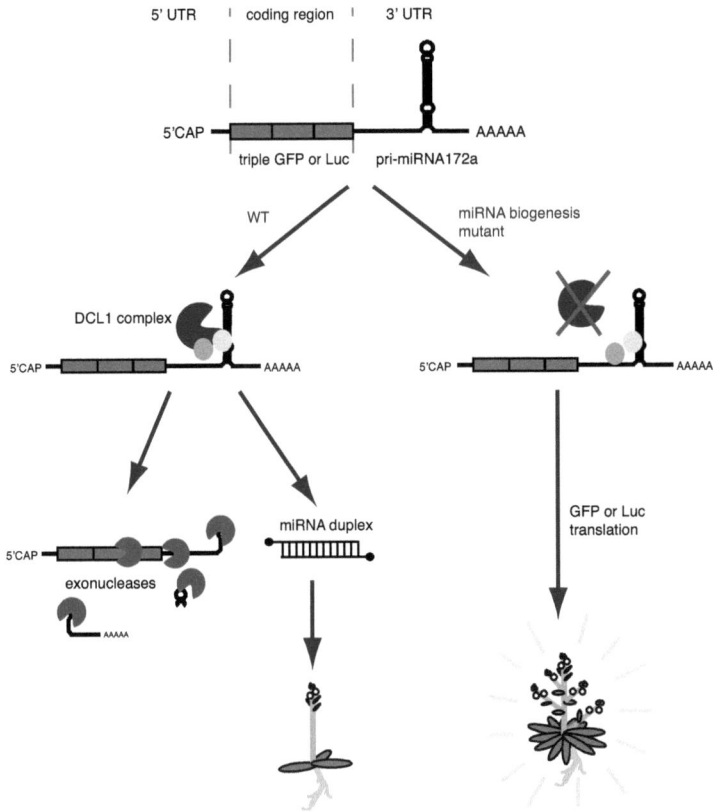

Figure 04: Hypothetical principles of 35S::Luc/triple GFP--pri-miR172a constructs
If the described construct is transformed into WT Col-0 plants, the miRNA duplex should be excised and the GFP or Luc mRNA degraded. The overexpressed miR172a should lead to an early flowering phenotype. In miRNA biogenesis mutants no processing takes place, the mRNA stays intact and the GFP or Luc can be translated, including a normal flowering phenotype.

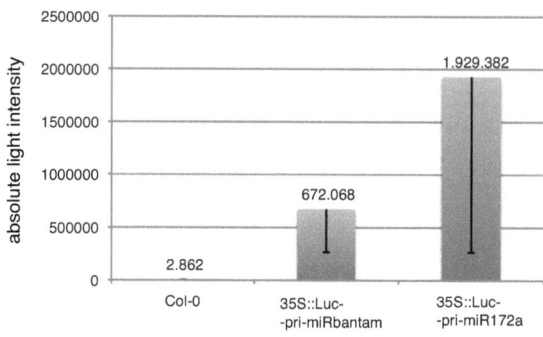

Figure 05: Luciferase activity in 35S::Luc—pri-miR172a transformed plants
Col-0 is the WT negative control, miRbantam the positive control. Error bars indicate standard deviation.

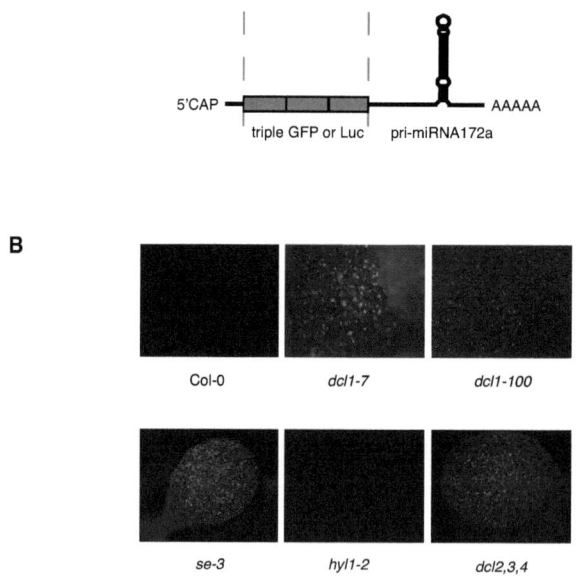

Figure 06: GFP activity in 35S::triple GFP—pri-miR172a transformed plants
(A) Schematic construct (B) GFP construct in different miRNA biogenesis mutants

DISCUSSION

I designed two reporter constructs overexpressing luciferase or triple GFP to monitor pri-miRNA processing in Arabidopsis. Together with pri-miR172a as 3'UTR I expected to obtain a read out of processing efficiency in WT or in miRNA biogenesis mutants, based on the reporter protein and the miR172a overexpressing phenotype. However, the actual results were difficult to reconcile with what we knew about miRNA processing.

The observation that the luciferase construct (35S::Luc—pri-miR172a) had a three times higher luciferase activity than its positive control is difficult to explain. Moreover, 35S::Luc—pri-miR172a overexpressor plants never showed a phenotype that was as strong as reported for 35S::pri-miR172a overexpressor plants, which exhibit a transformation of floral organs. These results suggest that the luciferase mRNA component somehow interfered with correct and/or efficient miRNA processing in this construct. I further tested a 35S::Luc—pri-miR319a construct using the overexpressor phenotype of miR319a (Palatnik et al., 2003) and compared it to 35S::pri-miR319a plants. The result was similar, which means that the Luc construct led to a much weaker overexpressor phenotype. It can be that sequences in the Luc mRNA interfere with processing of adjacent miRNA foldbacks: Either, by forcing the miRNA foldback into a conformation, which cannot be correctly recognised by DCL1, or by sterically preventing an interaction between the miRNA and DCL1. To avoid the apparent negative influence of the Luciferase sequence on miRNA processing, it would be interesting to introduce the pri-miRNA into the 5'UTR of the luciferase construct.

A GFP construct did work as expected in wild-type Col-0 plants. I could not detect a GFP signal above background levels and the plants showed a clear miR172a overexpression phenotype. Unfortunately, the results for this construct in miRNA biogenesis mutants (*dcl1, se, hyl1*) were not consistent with expectations. I expected in the mutants a GFP signal due to the impaired processing machinery and the unprocessed pri-miR172a. This was indeed the case for *dcl1* and *se* mutants but not for the *hyl1* mutant. It is possible, that SE plays a more important role in miRNA processing than HYL1, because null mutants of SE are embryonic lethal in contrast to *hyl1* mutants (Lu and Fedoroff, 2000; Prigge and Wagner, 2001). Therefore, miRNA processing in *se* mutants might be severely reduced, leading to a decreased amount of miRNAs. In contrast to *hyl1* mutants, which could have a functional but not accurate processing machinery. Here the miRNA foldback can be cleaved without giving rise to correct miRNAs, but random small RNAs.

Surprisingly I observed an early flowering phenotype but also a GFP signal when I transformed the GFP construct into *dcl2,3,4* mutants. This was unexpected because the miRNA processing machinery should not be effected by mutations in the *DCL2,3,4* genes. A possible explanation came from a paper published shortly after I obtained these results (Luo and Chen, 2007). The authors showed that truncated, unpolyadenylated transcripts are targeted by the RNA-DEPENDENT RNA POLYMERASE 6 (RDR6)/DCL4 silencing pathway. They used a triple GUS open reading frame for their experiments, which was very similar to our triple GFP. Thus the following scenario is likely: the miR172a in the 35S::triple GFP—pri-miR172a construct is processed by the DCL1 complex. The remaining, now truncated and unpolyadenylated 35S::triple GFP fragment is targeted by RDR6, amplified and processed into 24 nt small RNAs by DCL4. This would result in an early flowering phenotype and without a GFP signal. In a *dcl4* mutant background, the DCL1 dependent processing would be still functional but the degradation of the accumulated GFP 5' construct fragment by DCL4 would not take place and therefore the remaining GFP mRNA could be translated. Until recently it was thought that translational initiation could only take place on mRNAs that posses a 5' CAP structure as well as a polyA-tail. Especially the interaction of the 5' CAP and the polyA-binding proteins is needed for translation initiation (Gallie, 1991; Wells et al., 1998). But more and more studies focus on internal ribosomal entry sites (IRESs) (Jackson et al., 2010), facilitating a 5' CAP – polyA-tail independent translation possible. Although it is very unlikely, I cannot exclude this possibility for the RDR6-amplified GFP 5' fragments.

Since our initial construct did not behave as expected, I simplified the experimental design and overexpressed pri-miR172a alone, focussing on the phenotype and the miRNA level as a readout (see chapter III).

CHAPTER II

INTRODUCTION

Sequence specific nucleic acid binding proteins are known for a long time, with DNA binding transcription factors as the most prominent members among them. Transcription initiation/repression was thought to be the main factor in gene regulation and together with protein stability responsible for protein activity. However, transcription regulation cannot be seen as an ON/OFF switch. Studies have shown that for example chromatin modification can spread across neighbouring genes resulting in transcription leakage and low-grade spontaneous transcription occurs much more often than expected (Blake et al., 2003; Kaplan et al., 1992). Furthermore, transcription in certain cells does not need to fulfil the function of protein production, but could be necessary to respond faster to developmental signals (Rodriguez-Trelles et al., 2005; Yanai et al., 2006). It would not be necessary to change the chromatin status of the gene or to remove strong negative regulators if the gene has always a very low activity. However, it would be necessary to regulate the gene transcript on the next level, which is post-transcription. Therefore, the focus of research switched from transcription control to post-transcriptional events.

Seydoux and Braun suggested an RNA-centric program of post-transcriptional regulation, which would ensure the remaining plasticity of a genomic response (Seydoux and Braun, 2006). In such a scenario, RNA binding proteins (RBPs) and small RNAs co-ordinately regulate functionally related transcripts. These transcripts are called RNA-operons and could be easily regulated by the cell in response to environmental changes (Keene, 2007). The factors (RBPs and sRNAs) bind to multiple regulatory elements within specific mRNAs. These elements are called USERs (untranslated sequence elements for recognition). One mRNA can contain USERs for different RBPs or sRNAs and therefore be regulated by several factors.

At the beginning of my PhD thesis, it was not known whether such a sequence element did exist in miRNA precursors, perhaps less for regulation but more for the processing of miRNAs. A consensus sequence in the miRNA foldback could be recognized by a RBP and position the DCL1-complex directly or indirectly on the correct processing site. I analysed several miRNA precursors and compared the sequence in and between miRNA families to identify such USERs.

Another possible element for guiding processing could be the inner energy of miRNA precursor. Every long RNA in the cell forms secondary structures like loops and stems through hydrogen bonds. These folded RNA molecules are energetically favoured compared to the unfolded, single-stranded RNA. Each foldback has a specific energy profile depending on the secondary structure of the molecule. In theory this profile can be used by RBPs to recognize low energy regions spanning several base-pairs within a foldback (Han et al., 2006). It might be possible that the energy of a specific region in the foldback drops under a certain energy value for several base pairs. If these "low energy base pairs" are always a fixed number of nucleotides away from the miRNA duplex, they can be used as a landmark to guide the DCL1-complex onto the miRNA duplex. I used the web-based program *mfold* for modelling the secondary structure and the energy profile (Zuker, 2003) of miRNA foldbacks.

RESULTS

A common sequence binding motif in the miRNA precursors?

From other studies by colleagues in the lab I knew a sequence based recognition motif was unlikely to exist in the mature miRNA itself, because the miRNA can be exchanged against completely unrelated sequences (Schwab et al., 2006). Therefore, I analysed the flanking regions of the miRNA duplex, starting with a range of ten nucleotides (Figure 07B).

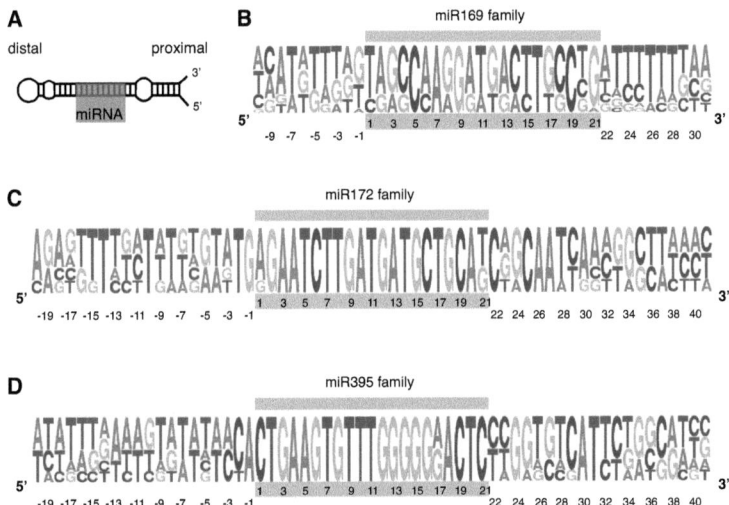

Figure 07: Sequence distribution around miRNAs
Relative frequency of bases in miRNA families, depending on their position relative to the miRNA sequence. First nucleotide (nt) of the miRNA is indicated with 1. Upstream nt with negative numbers, downstream nt with numbers > 21. MiRNA marked in light blue. **(A)** Schematic miRNA foldback **(B)** Graphical analysis of nt distribution in miR169 family members. This family contains two miRNA sequences, shifted by one nt to each other **(C)** and **(D)** Graphical analysis of nt distribution in miR172 family and miR395 family, respectively.

Later I extended the distance to twenty nucleotides (Figure 07C-D), because nearly all miRNA foldbacks contain shorter, but not longer flanking regions (e.g. miR159, miR172, miR390, miR393, etc.). I looked for a conserved sequence in a fixed distance to the proximal or distal end of the miRNA duplex, which could function as a USER. The visualisation of the analysis was done with Weblogo (Crooks et al., 2004; Schneider and Stephens, 1990). I could not find any sequence homology among analysed miRNA foldbacks of different families. Even within miRNA families, sequences outside the mature miRNA were rarely conserved (Figure 07). Especially families with several members did not show any conservation.

A landmark in the energy profile of miRNA foldbacks?

I also analysed the energy profile of miRNA foldback secondary structures (Figure 08) to look for a common energy landmark or footprint, which all or most of the miRNA foldbacks might possess. Such landmark could be used to position the DCL1-complex onto the miRNA duplex. An inner energy calculator from the web-based *mfold* program was used for the analysis.

I could not find any common landmark that occurred in all miRNA foldbacks. However, on average the lowest energy was at the position of the miRNA (Figure 09). For this analysis I used the sliding window method with a window size of 21 nt. MiRNAs are 21 nt in length and when the sliding window covered exactly the sequence of the miRNA it showed the lowest energy value. This means that at the position of the miRNA is the most stable region in the miRNA foldback, indicating potentially a significant role for miRNA processing. I could also show for the analysed foldback structure, that the region of the miRNA duplex contained always a longer stretch of very stable folding (ΔG lower than -2). This stable region was never at the same position in the miRNA duplex or showed the same characteristics. However, the overall inner energy of the analysed miRNA duplexes was lower compared to the surrounding energy profile (Figure 08B-G).

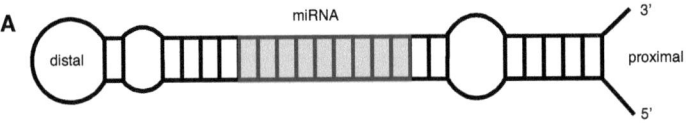

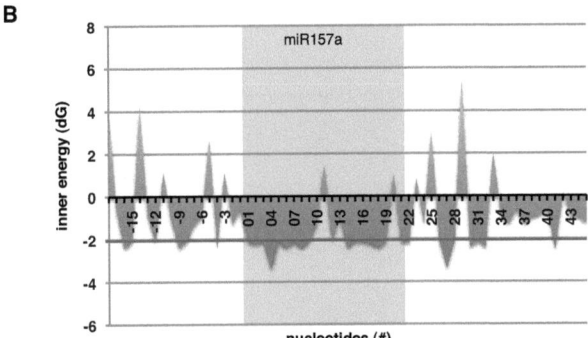

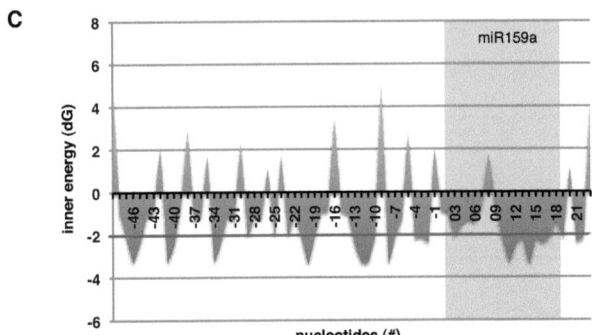

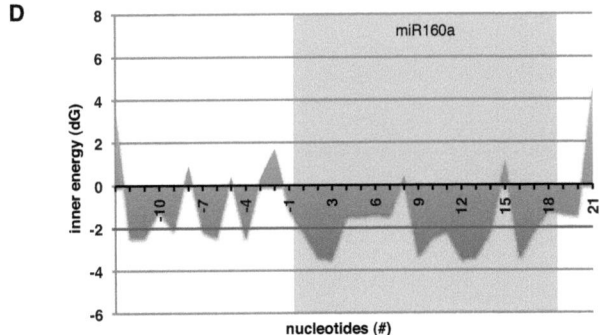

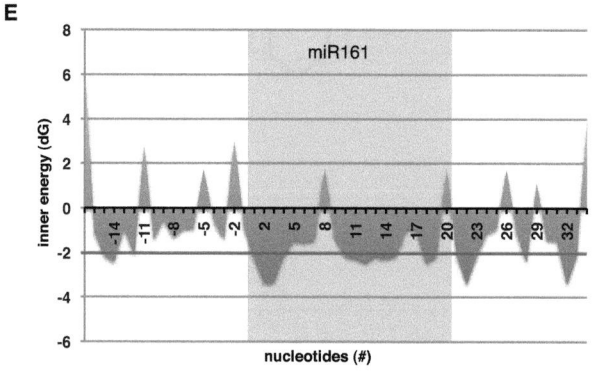

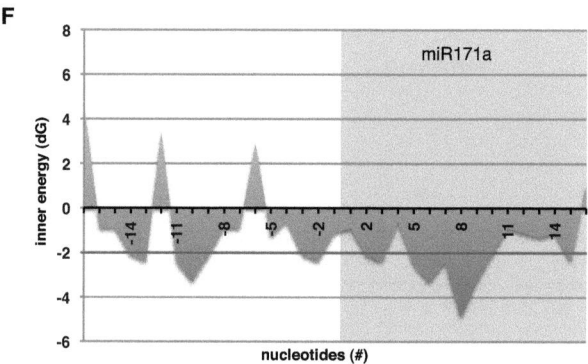

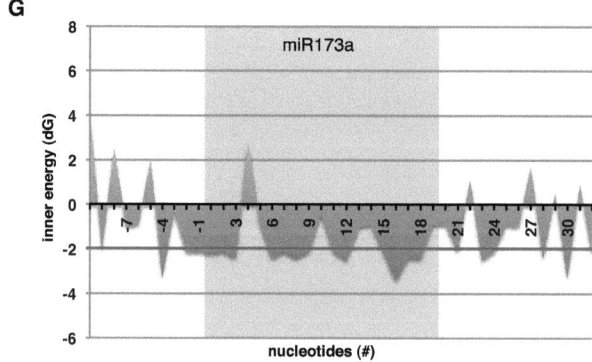

Figure 08: Energy profiles of miRNA secondary structures (previous page)
MiRNA position marked in grey **(A)** Schematic miRNA foldback **(B)** - **(G)** Energy landscape of miRNA foldback structures. The first nucleotide of the miRNA is indicated with 1. The terminal loop is always at the left side (see **A**). The red line indicates a threshold, which hypothetical has to be crossed for a longer stretch in the area of the miRNA sequence.

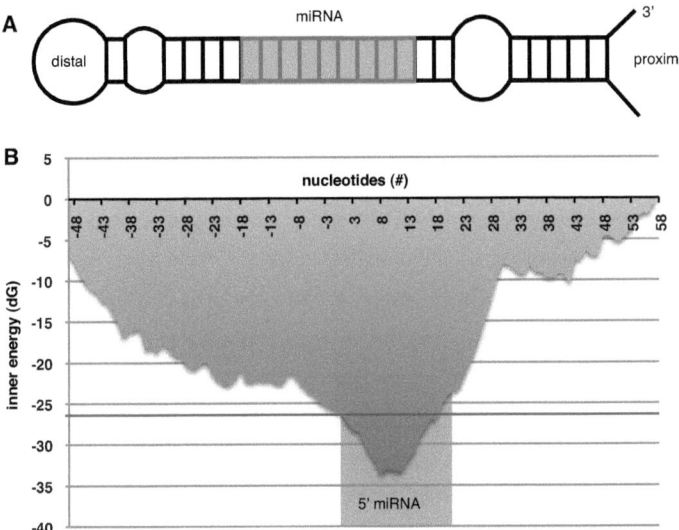

Figure 09: Average energy profile of miRNA secondary structure
MiRNA position marked in grey **(A)** Schematic miRNA foldback **(B)** Average energy landscape of miRNAs positioned at the 5' arm of the foldback, using the sliding window method (window 21 nt)

DISCUSSION

Analysis of the nucleotide sequence surrounding miRNA duplexes did not reveal any consensus motif conserved across miRNA families, and only very few conserved nucleotides in families itself. Taking the small size of such miRNA families into account (e.g. three or four members), this is likely due to chance similarity.

This analysis was made by hand for a few miRNA families and therefore I cannot exclude absolutely the presence of such a motif. Maybe a more sophisticated approach analysing all conserved miRNA families would give a different result, but it is unlikely that a strong signal would be found. Taking our data from chapter III into account, it seems clear that processing determinants are based on structure and not sequence.

The analysis of miRNA foldback energy profiles gave a different picture. In average the area of the miRNA duplex is very stable and contains the lowest ΔG values. Even the energy profiles of most analysed single miRNA foldbacks showed a longer stretch (5-10 nt) of stable folding (ΔG lower than -2) in the miRNA duplex. On the other hand, this stable stretch was never at the same position relative to the nucleotides of the miRNA duplex. And it never showed a similar profile. It is not likely that this feature is sufficient to place the DCL1-processing complex precisely onto the miRNA, but it is likely that it plays an important role in recognizing a miRNA containing foldback *per se*. In fact, every transcribed RNA forms secondary structure containing mainly stem loops or foldbacks. The unusual stable region in the pri-miRNA foldback could be used to distinguish between a miRNA containing and non-containing foldback.

CHAPTER III

INTRODUCTION

Since naturally occurring small point mutations have shed light on miRNA processing in humans (Duan et al., 2007; Sun et al., 2009), I performed mutagenesis of a miRNA precursor to identify sensitive sites in the miRNA foldback required for efficient processing. As an assay system, I used overexpression of miR172a, which causes very early flowering in transformants (Aukerman and Sakai, 2003; Chen, 2004; Park et al., 2002; Schwab et al., 2005). The usefulness of this assay was initially tested by my colleague Heike Wollmann with EMS (ethyl methanesulphonate) mutagenesis of miR172a overexpressing plants. Three lines, in which the early-flowering phenotype was partially suppressed, had mutations in the transgene (Figure S1c). Two of the mutations were in the miRNA itself, but one was 4 base pairs proximal to the miRNA/miRNA* duplex. Since it was outside the miRNA, it was likely to interfere with the accuracy or efficiency of processing, rather than ability of miR172a to reduce activity of its target genes, indicating the usefulness of mutagenizing the miR172a foldback.

With the detailed analysis of one miRNA foldback I sought to obtain information about determinants necessary for exact processing of this foldback. The information could be used to formulate general rules or identify necessary structural features for miRNA processing. With this data at hand I could check the majority of miRNA precursors, if their structure match our postulated rules.

RESULTS

Effects of point mutations on pri-miR172a processing efficiency

I decided to introduce a series of additional point mutations that either disrupted base pairing or closed unpaired bases in the foldback, in order to identify structural features important for miRNA processing. All point mutations were on the 5' arm of the miRNA foldback, opposite to the mature miRNA, to avoid confounding effects caused by reduced activity of miR172a itself. See Figure S3 for the effects of each mutation on foldback structure and the exact phenotype.

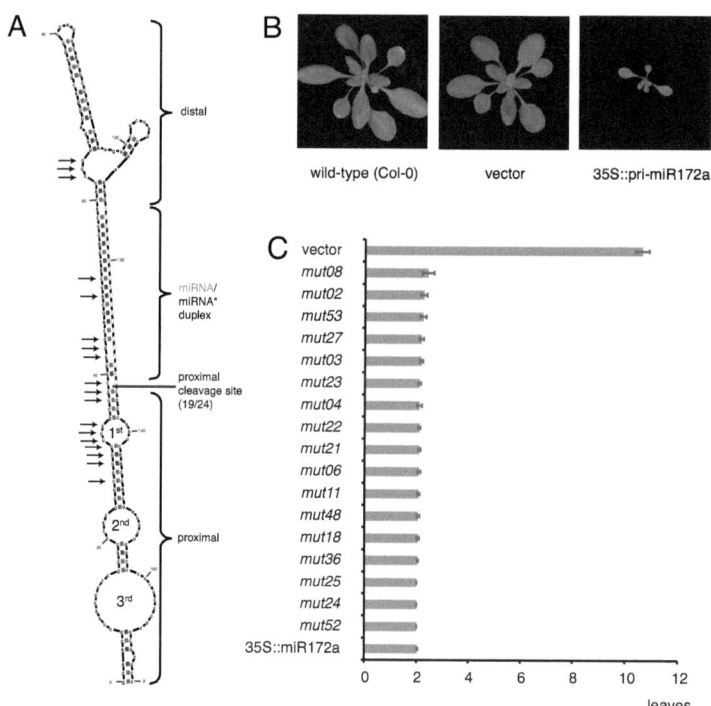

Figure 10: miR172a overexpression (previous page)
(A) Secondary structure of the miR172a foldback predicted with *mfold3.2* (Zuker, 2003). Arrows indicate the position of point mutations that do not affect the over-expression phenotype. "1st", "2nd" and "3rd" indicate the different proximal unpaired regions referred to in the text. The proximal cleavage site mapped by 5' RACE is indicated, with the fraction of corresponding clones given. **(B)** Phenotype of untransformed plants (Col-0), negative control transformed only with vector, and plants over-expressing miR172a from the wild-type precursor. **(C)** Flowering time of point mutants shown in (A), measured as number of leaves produced on the main stem before the first flower. Error bar indicates standard error of the mean. At least 15 T1 plants were analyzed for each construct.

Most point mutations did not alter pri-miR172a activity, as deduced from the early flowering phenotype of transgenic plants overexpressing the mutant constructs (Figure 10; Table S1; Figure S2 to S4). Several mutations were around the first, proximal processing site, showing a tolerance to minor structural changes at this position (e.g., *mut18*; Figure S3w).

Processing determinants in the proximal region of miR172a foldback

In the following, I divide the miRNA foldback into three parts: the proximal region, which contains three major unpaired regions between the base of the foldback and the miRNA/miRNA* duplex; the miRNA/miRNA* duplex itself; and the distal region with the terminal loop (Figure 10A). In animals, the proximal cleavage site of the DCL1-processing complex is determined by the distance from the miRNA/miRNA* duplex to the base of the foldback, which constitutes the 5' and 3' ends of the pre-miRNA (Han et al., 2006). In plants, in contrast, the distance from the miRNA/miRNA* duplex to the base of the foldback is highly variable (Fahlgren et al., 2007).

I introduced several deletions to determine the importance of features within the foldback base (Figure 11). In the pri-miR172a foldback are three unpaired regions proximal to the miRNA/miRNA* duplex. Deletion of the third and largest proximal unpaired region, or the second and third proximal unpaired regions did not compromise pri-miR172a activity (mutants *stem1* and *stem2*; Figure 11; Figure S3c, S3d). On the contrary, deletion of the proximal portion including the first unpaired region abolished the ability of pri-miR172a to cause early flowering (mutant *stemcore*; Figure 11; Figure S3e). The phenotypic effects of the mutant transgenes were closely paralleled by the amount of miR172a detected on small RNA blots (Figure 11B).

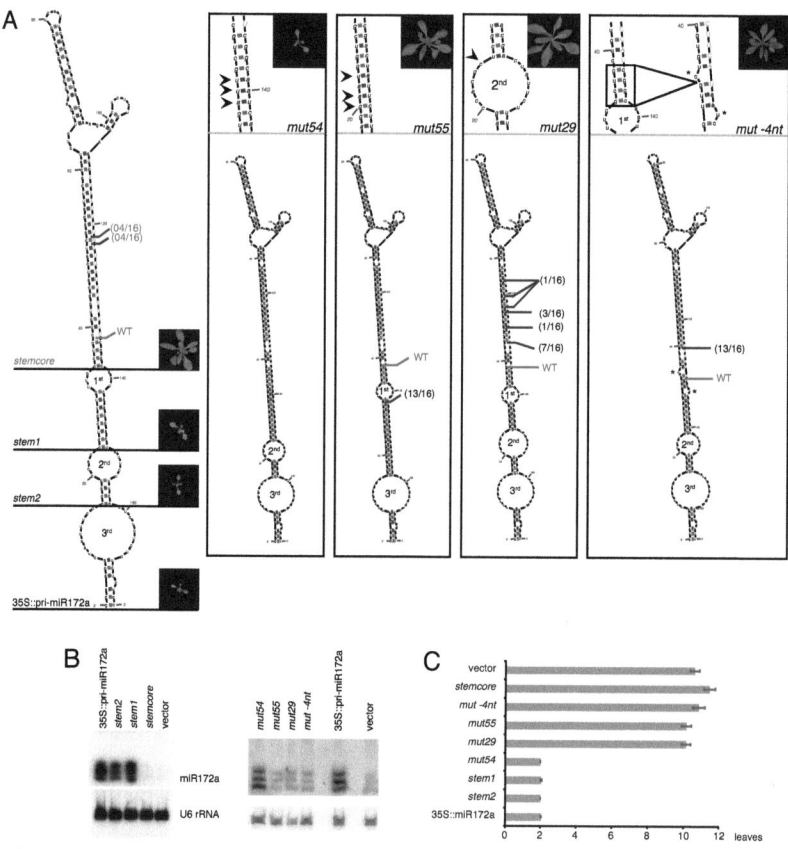

Figure 11: Analysis of the proximal miR172a foldback region.
(A) Predicted secondary structures and phenotypes of mutants. Stemcore, stem1 and stem2 are mutants in which the regions below the horizontal lines were deleted. Mapped cleavage sites of mutants are indicated with blue lines, with the fraction of clones corresponding to each position given. The wild-type proximal cleavage site is indicated in grey. Arrowheads indicate mutated bases in mut54, mut55 and mut29; the small black frame indicates a deletion in mut -4nt. Black stars indicate asymmetrical, unpaired bases. (B) Small RNA blots. RNAs extracted from three biological replicates were loaded consecutively in each lane. As loading control, U6 rRNA was used. Because of its larger size, the replicates are not clearly resolved for U6. (C) Flowering time of mutants. Error bar indicates standard error of the mean. At least 15 T1 plants were analyzed for each construct.

To investigate the effect of the *stemcore* mutation on miR172a processing, I performed 5'-rapid amplification of cDNA ends (5' RACE) to map the proximal, and therefore 3' most, cleavage site of the DCL1 processor complex in the pri-miR172a foldback (Figure S2e) (Llave et al., 2002).

Compared to wild-type pri-miR172a, the cleavage sites were shifted 12 to 13 nucleotides distally (Figure 10A, O4A), which was 16 to 17 base pairs from the base of the *stemcore* stem loop. An intermediate cleavage site, proximal to the first processing site (Kurihara and Watanabe, 2004), was not detected.

Concentrating on this processing-sensitive region, I introduced more subtle mutations proximally to the first unpaired region. Closing the first unpaired region (*mut54*) had no effect, consistent with results obtained with point mutations in this region (Figure S3ba). Closing the second unpaired region, however, led to a loss of miR172a product in *mut55*, due to a proximal shift of the first processing site by 8 nucleotides (Figure 11A, B). Notably, the new processing site is now 15 nucleotides distal from the third unpaired region.

A very interesting point mutant was *mut29*, in which the second unpaired region was enlarged at its distal end and the stem separating the first and second unpaired regions was correspondingly shortened (Figure 11; Figure S3ah). In *mut29*, the main cleavage site was not only shifted distally, but the accuracy of processing was also compromised. Together with the results from mutant *stem1*, this observation shows that it is not the unpaired region *per se*, but the transition from single-stranded RNA to double-stranded RNA that is required for positioning the DCL1-complex 14 to 15 base pairs proximal to the miRNA/miRNA* duplex (see also *mut30*; Figure S3ai). The intervening double-stranded stem of at least 6 base pairs can tolerate small 1-nucleotide bulges (see *mut24*, *mut25*, *mut26*, *mut27* in Figure 10 and Figure S3ac-af).

All proximal mutants that retained the 15 base pair distance between the second unpaired region and the miRNA/miRNA* were able to induce very early flowering. In further support of the 15-base pair rule, deleting 4 base pairs of the stem just proximal to the miRNA/miRNA* in *mut -4nt* caused a distal shift of the processing site (Figure 11). Importantly, the shift of the cleavage site was 5, not 4, nucleotides. The shift by one extra base pair was apparently caused by the unpaired bases (indicated by asterisks in Figure 11a) proximately to the cleavage site. It seems that asymmetrical, unpaired nucleotides do not contribute to the measured distance. I propose that this is due to the covalent bond on the opposite side of the molecule, which cannot be stretched.

Processing determinants in the distal region of miR172a foldback

After identifying structural features of the proximal region important for precise miRNA processing, I investigated the contribution of the distal portion of the miR172a foldback. Minor changes of the structure of the distal region, such as pairing the single stranded first nucleotide of the miRNA in *mut01* (Figure 12A), had little effect on the ability of pri-miR172a to induce early flowering. Even deleting almost the entire distal portion in *mut63*, leaving only a single paired base pair beyond the miRNA/miRNA* and a predicted 4 nucleotide terminal loop, did not completely abolish the miR172a over-expression phenotype (Figure 12; Figure S3bg). Somewhat surprisingly, stronger effects were seen when I disrupted the defined stem-like secondary structure at the distal processing site of the miRNA/miRNA* duplex itself, in *mut14* and *mut15*. Opening this region strongly reduced the levels of overexpressed miR172a (Figure 12; Figure S3s, S3t). The correct proximal cleavage site was, however, unaffected. In addition to the miRNA, a larger RNA species of about 50 bases is seen, which likely corresponds to the pre-miRNA, i.e., the miRNA/miRNA* duplex including the much shortened terminal loop (Figure 12B).

Design of a minimal miRNA

Based on the insights gleaned from the mutants, I designed a minimal pri-miRNA construct called pri-amiR-CH42 (Figure 13A). The mature amiR-CH42 sequence in the pri-miRNA construct was adopted from an artificial miRNA (amiRNA) (Schwab et al., 2006), that targets At4G18480 (*CH42*), a gene involved in chlorophyll biosynthesis (Koncz et al., 1990). Knock-down of *CH42* with this amiRNA leads to an easily recognized bleaching phenotype (Felippes and Weigel, 2009). The structure of the miRNA duplex contributed very little to correct processing, as long as the duplex was in a defined stem-like shape (Figures 03 and 05, and Figure S3). Therefore I copied the pattern of paired and unpaired bases of the miRNA/miRNA* duplex for amiR-CH42 from miR172a.

The non-miRNA/miRNA* portion of the foldback was designed *de novo* based on the information gleaned from the mutant analyses. The proximal portion of the foldback, which was kept as short as possible, contained a five nucleotide unpaired region and a stem of 14 base pairs of random sequence, which ended with G:C base pairs before the unpaired region, to maximise stability. This structural feature was adopted, because several miRNA foldbacks including miR172a have a partially paired stem of 14 to 17 base pairs proximal to the miRNA/miRNA* duplex (Figure 13B). A systematic survey of all *A. thaliana* miRNAs conserved in *Populus trichocarpa* (Fahlgren et al., 2007), excluding the related miR159 and miR319 miRNA, which are processed by a distinct loop-to-base mechanism (Palatnik et al., 2007) confirmed that unpaired regions proximal to the miRNA/miRNA* duplex increase substantially in size beyond a point that is about 15 nucleotides from the duplex (Figure 13C). Since *mut63* had suggested that too short a distal stem reduced processing efficiency (Figure 12; Figure S3bg), a seven base pair stem was placed distal to the miRNA/miRNA* duplex of pri-amiR-CH42. To disentangle direct effects of amiR-CH42 on its target from those caused potentially by amplification and secondary small RNAs, pri-amiR-CH42 was expressed under the *SUC2* (At1G22710) promoter, which is active only in phloem companion cells (Imlau et al., 1999).

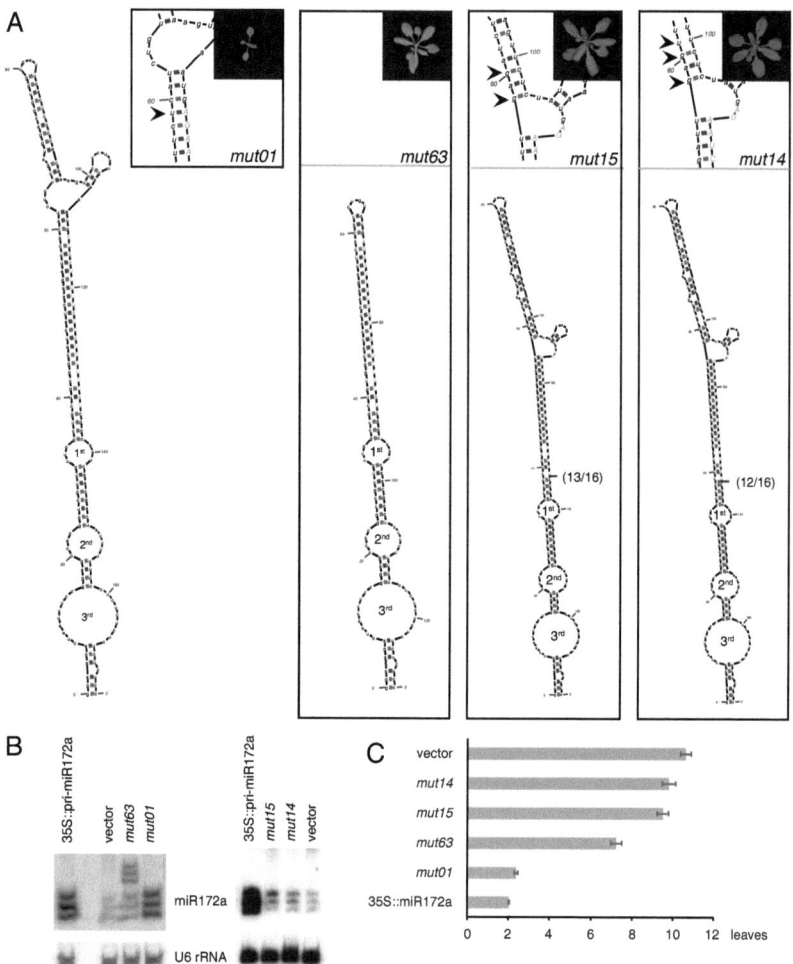

Figure 12: Analysis of the distal miR172a foldback region.
(A) For comparison, the wild-type miR172a foldback is shown on the left, insets show details of mutants. **(B)** Small RNA blots. **(C)** Flowering time of mutants. Error bar indicates standard error of the mean. At least 15 T1 plants were analyzed for each construct.

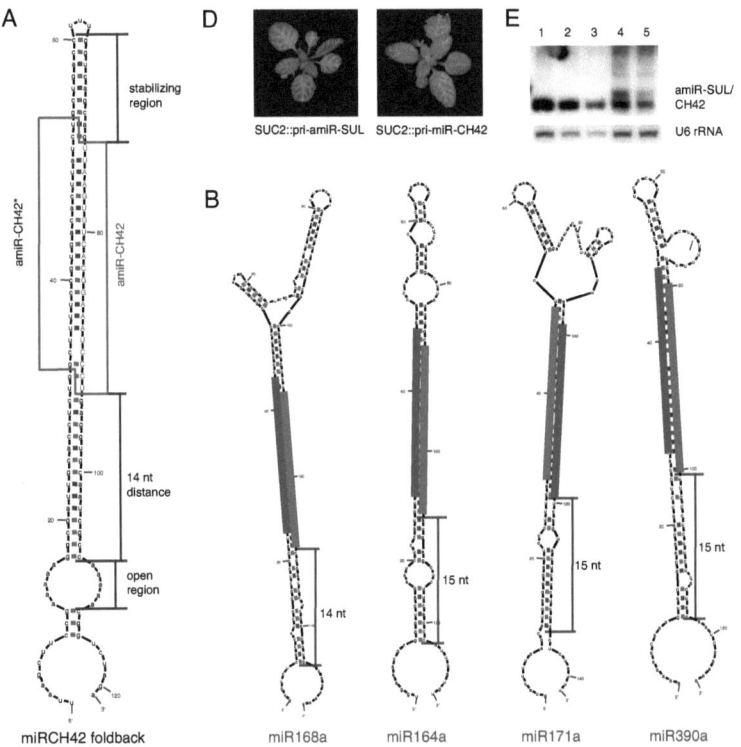
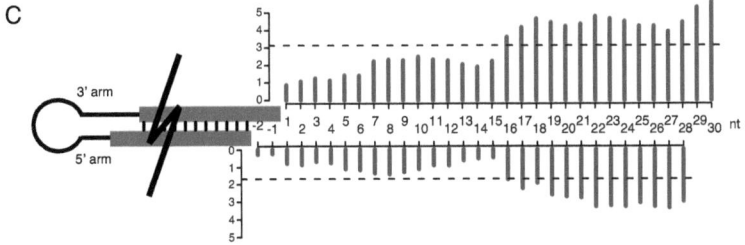

Figure 13: Minimal amiR-CH42 and secondary structures of miRNA foldbacks (previous page)
(A) Predicted secondary structure of pri-amiR-CH42. **(B)** Examples of miRNA foldbacks with a similar structure as pri-miR172a, including a 14 to 15 nucleotide stem proximal to the miRNA/miRNA* duplex. The miRNA is indicated in blue, the miRNA* in gray. **(C)** Position weight matrix analysis for all A. thaliana miRNAs conserved in P. trichocarpa. X-axis indicates distance from miRNA/miRNA* duplex (green boxes). Y-axes indicate average run of unpaired bases. Dashed lines indicate the overall average of unpaired bases. From 15 nucleotides distal to the miRNA/miRNA* duplex, the size of unpaired regions is above the average (in red). **(D)** Representative plants expressing amiR-SUL, produced from the natural miR319a backbone, and amiR-CH42, processed from an entirely artificial precursor. **(E)** Small RNA blots. The first three lanes were loaded with 4, 2, and 1 µg of RNA from a SUC2::pri-amiR-SUL line, and the last two lanes with 4 µg of RNA each from two different SUC2::pri-amiR-CH42 T1 plants.

Plants expressing pri-amiR-CH42 showed a similar, but somewhat weaker bleaching phenotype around leaf veins (Figure 13D) compared to controls expressing an amiRNA against *CH42* from the standard miR319a backbone (Schwab et al., 2006). A small RNA blot confirmed that the predicted amiR-CH42 was produced (Figure 13E). A second, larger species was also detected. This 24 nt small RNA species shows that the pri-amiR-CH42 was not only detected by the miRNA processing complex, but also by the siRNA processing complex. Therefore, the construct still can be improved to avoid the DCL3 siRNA pathway.

DISCUSSION

Animal mRNA foldbacks are usually ~33 base pairs long, with a proximal 11 base pair stem, the 21 base pair miRNA/miRNA* duplex, and a terminal loop at the distal end (reviewed in Winter et al., 2009). In plants, the miRNA-containing foldbacks are much more diverse in length and structure, with the miRNA/miRNA* duplex being at variable positions, including being very close to the proximal base of the foldback, such as in pri-miR159 and pri-miR319 (Axtell and Bartel, 2005; Qi et al., 2005; Warthmann et al., 2008). This is consistent with there being more than one processing pathways in Arabidopsis. It would be interesting to determine how they are distinguished on the level of processing protein-complexes and on the level of miRNA foldbacks.

Despite the complexity of miRNA processing in Arabidopsis, our extensive analysis on the miR172a foldback revealed essential secondary structural features of one of the processing pathways. Most point mutations did not alter pri-miR172a activity, as deduced from the early flowering phenotype of transgenic plants overexpressing the mutant constructs (Figure 10; Table S1; Figure S2 to S4). Several mutations were around the first, proximal processing site, indicating a tolerance to minor structural changes at this position (e.g., *mut18*; Figure S3w). In humans, a single nucleotide polymorphism in the miRNA duplex can strongly attenuate processing of pri-miRNAs to pre-miRNAs (Duan et al., 2007), suggesting that the Arabidopsis processing machinery is more tolerant to structural changes inside the miRNA duplex than its animal counterpart. On the contrary, the distal processing site appears to be quite sensitive, and needs to have a defined structure. In addition I could show, that distally to the miRNA/miRNA* duplex, a certain minimum length of the stem is important for accurate and efficient processing.

The most sensitive region to mutations is 15 nt proximal from the first processing site. A single-to-double stranded transition followed by an eight nucleotide stem is necessary for accurate miRNA processing. Also, several pri-miRNAs, such as pri-miR168a, -miR164a, -miR171a and -miR390a, share with pri-miR172a a transition between a major unpaired region and a paired stem 14 to 17 base pairs proximal to the miRNA/miRNA* duplex (Figure 13D, E). This transition is also conserved in the pri-miR172a foldback of *Arabidopsis lyrata* (Figure S4). Both the structure of endogenous pri-miRNAs and some of our pri-miR172a mutants indicate that the 14 to 17 base pair stem proximal to the miRNA/miRNA* duplex tolerates small unpaired bulges or similar variants, as long as the linear structure of the foldback is maintained. Equally, the extent of pairing in the miRNA/miRNA* duplex is flexible.

Based on these observations I could design *de novo* an artificial minimal miRNA construct (amiR-CH42), which is recognized as a miRNA foldback. This miRNA backbone gives rise to 21 nt small RNAs but also 24 nt RNAs. Several Arabidopsis miRNA genes have recently been shown to give rise to miRNAs of 23 to 25 nucleotide length in a DCL3-dependent fashion, in addition to the canonical 21 nucleotide species produced by DCL1, and this appears to be correlated with the length and extent of mismatches in the precursor (Vazquez et al., 2008). The amiR-CH42 foldback contains two long, perfectly paired stems of 14 and 15 base pairs, which is unusual for evolutionarily old miRNAs. Introducing mismatches in the stems could likely reduce the production of longer miRNAs.

In summary, I have identified structural features required for accurate and efficient processing of miR172a, and these appear to apply to the majority of Arabidopsis miRNAs, with the notable exception of miR159 and miR319 (Addo-Quaye et al., 2009; Bologna et al., 2009). It will be interesting to determine whether there are additional alternative pathways for miRNA processing in plants, and whether they all evolved at the same time.

REFERENCES

Addo-Quaye, C., Snyder, J.A., Park, Y.B., Li, Y.F., Sunkar, R., and Axtell, M.J. (2009). Sliced microRNA targets and precise loop-first processing of MIR319 hairpins revealed by analysis of the Physcomitrella patens degradome. Rna.

Allen, E., Xie, Z., Gustafson, A.M., and Carrington, J.C. (2005). microRNA-directed phasing during trans-acting siRNA biogenesis in plants. Cell *121*, 207-221.

Aukerman, M.J., and Sakai, H. (2003). Regulation of flowering time and floral organ identity by a MicroRNA and its APETALA2-like target genes. Plant Cell *15*, 2730-2741.

Axtell, M.J., and Bartel, D.P. (2005). Antiquity of microRNAs and their targets in land plants. Plant Cell *17*, 1658-1673.

Baumberger, N., and Baulcombe, D.C. (2005). Arabidopsis ARGONAUTE1 is an RNA Slicer that selectively recruits microRNAs and short interfering RNAs. Proc Natl Acad Sci USA *102*, 11928-11933.

Bernstein, E., Caudy, A.A., Hammond, S.M., and Hannon, G.J. (2001). Role for a bidentate ribonuclease in the initiation step of RNA interference. Nature *409*, 363-366.

Blake, W.J., M, K.A., Cantor, C.R., and Collins, J.J. (2003). Noise in eukaryotic gene expression. Nature *422*, 633-637.

Bologna, N.G., Mateos, J.L., Bresso, E.G., and Palatnik, J.F. (2009). A loop-to-base processing mechanism underlies the biogenesis of plant microRNAs miR319 and miR159. EMBO J.

Borsani, O., Zhu, J., Verslues, P.E., Sunkar, R., and Zhu, J.K. (2005). Endogenous siRNAs derived from a pair of natural cis-antisense transcripts regulate salt tolerance in Arabidopsis. Cell *123*, 1279-1291.

Brodersen, P., Sakvarelidze-Achard, L., Bruun-Rasmussen, M., Dunoyer, P., Yamamoto, Y.Y., Sieburth, L., and Voinnet, O. (2008). Widespread translational inhibition by plant miRNAs and siRNAs. Science *320*, 1185-1190.

Chen, X. (2004). A microRNA as a translational repressor of APETALA2 in Arabidopsis flower development. Science *303*, 2022-2025.

Clough, S.J., and Bent, A.F. (1998). Floral dip: a simplified method for Agrobacterium-mediated transformation of Arabidopsis thaliana. Plant J *16*, 735-743.

References

Cogoni, C., Irelan, J.T., Schumacher, M., Schmidhauser, T.J., Selker, E.U., and Macino, G. (1996). Transgene silencing of the al-1 gene in vegetative cells of Neurospora is mediated by a cytoplasmic effector and does not depend on DNA-DNA interactions or DNA methylation. EMBO J *15*, 3153-3163.

Crooks, G.E., Hon, G., Chandonia, J.M., and Brenner, S.E. (2004). WebLogo: a sequence logo generator. Genome Res *14*, 1188-1190.

Dong, Z., Han, M.H., and Fedoroff, N. (2008). The RNA-binding proteins HYL1 and SE promote accurate in vitro processing of pri-miRNA by DCL1. Proc Natl Acad Sci USA *105*, 9970-9975.

Duan, R., Pak, C., and Jin, P. (2007). Single nucleotide polymorphism associated with mature miR-125a alters the processing of pri-miRNA. Hum Mol Genet *16*, 1124-1131.

Elbashir, S.M., Lendeckel, W., and Tuschl, T. (2001). RNA interference is mediated by 21- and 22-nucleotide RNAs. Genes Dev *15*, 188-200.

Fahlgren, N., Howell, M.D., Kasschau, K.D., Chapman, E.J., Sullivan, C.M., Cumbie, J.S., Givan, S.A., Law, T.F., Grant, S.R., Dangl, J.L., et al. (2007). High-Throughput Sequencing of Arabidopsis microRNAs: Evidence for Frequent Birth and Death of MIRNA Genes. PLoS ONE *2*, e219.

Fang, Y., and Spector, D.L. (2007). Identification of nuclear dicing bodies containing proteins for microRNA biogenesis in living Arabidopsis plants. Curr Biol *17*, 818-823.

Felippes, F.F., and Weigel, D. (2009). Triggering the formation of tasiRNAs in Arabidopsis thaliana: the role of microRNA miR173. EMBO Rep *10*, 264-270.

Fire, A., Albertson, D., Harrison, S.W., and Moerman, D.G. (1991). Production of antisense RNA leads to effective and specific inhibition of gene expression in C. elegans muscle. Development *113*, 503-514.

Fire, A., Xu, S., Montgomery, M.K., Kostas, S.A., Driver, S.E., and Mello, C.C. (1998). Potent and specific genetic interference by double-stranded RNA in Caenorhabditis elegans. Nature *391*, 806-811.

Fujioka, Y., Utsumi, M., Ohba, Y., and Watanabe, Y. (2007). Location of a possible miRNA processing site in SmD3/SmB nuclear bodies in Arabidopsis. Plant Cell Physiol *48*, 1243-1253.

Gallie, D.R. (1991). The cap and poly(A) tail function synergistically to regulate mRNA translational efficiency. Genes Dev *5*, 2108-2116.

References

Gandikota, M., Birkenbihl, R.P., Hohmann, S., Cardon, G.H., Saedler, H., and Huijser, P. (2007). The miRNA156/157 recognition element in the 3' UTR of the Arabidopsis SBP box gene SPL3 prevents early flowering by translational inhibition in seedlings. Plant J.

Gregory, B.D., O'Malley, R.C., Lister, R., Urich, M.A., Tonti-Filippini, J., Chen, H., Millar, A.H., and Ecker, J.R. (2008). A link between RNA metabolism and silencing affecting Arabidopsis development. Dev Cell *14*, 854-866.

Grigg, S.P., Canales, C., Hay, A., and Tsiantis, M. (2005). SERRATE coordinates shoot meristem function and leaf axial patterning in Arabidopsis. Nature *437*, 1022-1026.

Gy, I., Gasciolli, V., Lauressergues, D., Morel, J.B., Gombert, J., Proux, F., Proux, C., Vaucheret, H., and Mallory, A.C. (2007). Arabidopsis FIERY1, XRN2, and XRN3 are endogenous RNA silencing suppressors. Plant Cell *19*, 3451-3461.

Ha, I., Wightman, B., and Ruvkun, G. (1996). A bulged lin-4/lin-14 RNA duplex is sufficient for Caenorhabditis elegans lin-14 temporal gradient formation. Genes Dev *10*, 3041-3050.

Hamilton, A.J., and Baulcombe, D.C. (1999). A species of small antisense RNA in posttranscriptional gene silencing in plants. Science *286*, 950-952.

Hammond, S.M., Bernstein, E., Beach, D., and Hannon, G.J. (2000). An RNA-directed nuclease mediates post-transcriptional gene silencing in Drosophila cells. Nature *404*, 293-296.

Han, J., Lee, Y., Yeom, K.H., Nam, J.W., Heo, I., Rhee, J.K., Sohn, S.Y., Cho, Y., Zhang, B.T., and Kim, V.N. (2006). Molecular Basis for the Recognition of Primary microRNAs by the Drosha-DGCR8 Complex. Cell *125*, 887-901.

Han, M.H., Goud, S., Song, L., and Fedoroff, N. (2004). The Arabidopsis double-stranded RNA-binding protein HYL1 plays a role in microRNA-mediated gene regulation. Proc Natl Acad Sci USA *101*, 1093-1098.

Hellens, R.P., Edwards, E.A., Leyland, N.R., Bean, S., and Mullineaux, P.M. (2000). pGreen: a versatile and flexible binary Ti vector for Agrobacterium-mediated plant transformation. Plant molecular biology *42*, 819-832.

Herr, A.J., Jensen, M.B., Dalmay, T., and Baulcombe, D.C. (2005). RNA polymerase IV directs silencing of endogenous DNA. Science *308*, 118-120.

Hofacker, I.L., Fontana, W., Stadler, P.F., Bonhoeffer, L.S., Tacker, M., and Schuster, P. (1994). Fast folding and comparison of RNA secondary structures. Monatsh Chem *125*, 167-188.

References

Howell, M.D., Fahlgren, N., Chapman, E.J., Cumbie, J.S., Sullivan, C.M., Givan, S.A., Kasschau, K.D., and Carrington, J.C. (2007). Genome-wide analysis of the RNA-DEPENDENT RNA POLYMERASE6/DICER-LIKE4 pathway in Arabidopsis reveals dependency on miRNA- and tasiRNA-directed targeting. Plant Cell *19*, 926-942.

Huang, Y., Ji, L., Huang, Q., Vassylyev, D.G., Chen, X., and Ma, J.B. (2009). Structural insights into mechanisms of the small RNA methyltransferase HEN1. Nature *461*, 823-827.

Imlau, A., Truernit, E., and Sauer, N. (1999). Cell-to-cell and long-distance trafficking of the green fluorescent protein in the phloem and symplastic unloading of the protein into sink tissues. Plant Cell *11*, 309-322.

Izant, J.G., and Weintraub, H. (1984). Inhibition of thymidine kinase gene expression by anti-sense RNA: a molecular approach to genetic analysis. Cell *36*, 1007-1015.

Jackson, R.J., Hellen, C.U., and Pestova, T.V. (2010). The mechanism of eukaryotic translation initiation and principles of its regulation. Nat Rev Mol Cell Biol *11*, 113-127.

Jamalkandi, S.A., and Masoudi-Nejad, A. (2009). Reconstruction of Arabidopsis thaliana fully integrated small RNA pathway. Funct Integr Genomics *9*, 419-432.

Kaplan, J.C., Kahn, A., and Chelly, J. (1992). Illegitimate transcription: its use in the study of inherited disease. Hum Mutat *1*, 357-360.

Keene, J.D. (2007). RNA regulons: coordination of post-transcriptional events. Nat Rev Genet *8*, 533-543.

Kim, S., Yang, J.Y., Xu, J., Jang, I.C., Prigge, M.J., and Chua, N.H. (2008). Two CAP BINDING PROTEINS CBP20 and CBP80 are involved in processing primary microRNAs. Plant Cell Physiol.

Koncz, C., Mayerhofer, R., Koncz-Kalman, Z., Nawrath, C., Reiss, B., Redei, G.P., and Schell, J. (1990). Isolation of a gene encoding a novel chloroplast protein by T-DNA tagging in Arabidopsis thaliana. EMBO J *9*, 1337-1346.

Kurihara, Y., Takashi, Y., and Watanabe, Y. (2006). The interaction between DCL1 and HYL1 is important for efficient and precise processing of pri-miRNA in plant microRNA biogenesis. Rna *12*, 206-212.

Kurihara, Y., and Watanabe, Y. (2004). Arabidopsis micro-RNA biogenesis through Dicer-like 1 protein functions. Proc Natl Acad Sci USA *101*, 12753-12758.

References

Lagos-Quintana, M., Rauhut, R., Lendeckel, W., and Tuschl, T. (2001). Identification of novel genes coding for small expressed RNAs. Science *294*, 853-858.

Lau, N.C., Lim, L.P., Weinstein, E.G., and Bartel, D.P. (2001). An abundant class of tiny RNAs with probable regulatory roles in Caenorhabditis elegans. Science *294*, 858-862.

Laubinger, S., Sachsenberg, T., Zeller, G., Busch, W., Lohmann, J.U., Ratsch, G., and Weigel, D. (2008). Dual roles of the nuclear cap-binding complex and SERRATE in pre-mRNA splicing and microRNA processing in Arabidopsis thaliana. Proc Natl Acad Sci USA *105*, 8795-8800.

Lee, R.C., and Ambros, V. (2001). An extensive class of small RNAs in Caenorhabditis elegans. Science *294*, 862-864.

Lee, R.C., Feinbaum, R.L., and Ambros, V. (1993). The C. elegans heterochronic gene lin-4 encodes small RNAs with antisense complementarity to lin-14. Cell *75*, 843-854.

Lee, Y., Kim, M., Han, J., Yeom, K.H., Lee, S., Baek, S.H., and Kim, V.N. (2004). MicroRNA genes are transcribed by RNA polymerase II. EMBO J *23*, 4051-4060.

Li, J., Yang, Z., Yu, B., Liu, J., and Chen, X. (2005). Methylation protects miRNAs and siRNAs from a 3'-end uridylation activity in Arabidopsis. Curr Biol *15*, 1501-1507.

Liu, P.P., Montgomery, T.A., Fahlgren, N., Kasschau, K.D., Nonogaki, H., and Carrington, J.C. (2007). Repression of AUXIN RESPONSE FACTOR10 by microRNA160 is critical for seed germination and post-germination stages. Plant J *52*, 133-146.

Llave, C., Xie, Z., Kasschau, K.D., and Carrington, J.C. (2002). Cleavage of Scarecrow-like mRNA targets directed by a class of Arabidopsis miRNA. Science *297*, 2053-2056.

Lobbes, D., Rallapalli, G., Schmidt, D.D., Martin, C., and Clarke, J. (2006). SERRATE: a new player on the plant microRNA scene. EMBO Rep *7*, 1052-1058.

Lu, C., and Fedoroff, N. (2000). A mutation in the Arabidopsis HYL1 gene encoding a dsRNA binding protein affects responses to abscisic acid, auxin, and cytokinin. Plant Cell *12*, 2351-2366.

Luo, Z., and Chen, Z. (2007). Improperly terminated, unpolyadenylated mRNA of sense transgenes is targeted by RDR6-mediated RNA silencing in Arabidopsis. Plant Cell *19*, 943-958.

MacRae, I.J., Zhou, K., and Doudna, J.A. (2007). Structural determinants of RNA recognition and cleavage by Dicer. Nat Struct Mol Biol *14*, 934-940.

Macrae, I.J., Zhou, K., Li, F., Repic, A., Brooks, A.N., Cande, W.Z., Adams, P.D., and Doudna, J.A. (2006). Structural basis for double-stranded RNA processing by Dicer. Science *311*, 195-198.

References

Margis, R., Fusaro, A.F., Smith, N.A., Curtin, S.J., Watson, J.M., Finnegan, E.J., and Waterhouse, P.M. (2006). The evolution and diversification of Dicers in plants. FEBS Lett *580*, 2442-2450.

Matzke, M.A., Primig, M., Trnovsky, J., and Matzke, A.J. (1989). Reversible methylation and inactivation of marker genes in sequentially transformed tobacco plants. EMBO J *8*, 643-649.

Mi, S., Cai, T., Hu, Y., Chen, Y., Hodges, E., Ni, F., Wu, L., Li, S., Zhou, H., Long, C., *et al.* (2008). Sorting of small RNAs into Arabidopsis argonaute complexes is directed by the 5' terminal nucleotide. Cell *133*, 116-127.

Montgomery, T.A., Howell, M.D., Cuperus, J.T., Li, D., Hansen, J.E., Alexander, A.L., Chapman, E.J., Fahlgren, N., Allen, E., and Carrington, J.C. (2008). Specificity of ARGONAUTE7-miR390 interaction and dual functionality in TAS3 trans-acting siRNA formation. Cell *133*, 128-141.

Moss, E.G., Lee, R.C., and Ambros, V. (1997). The cold shock domain protein LIN-28 controls developmental timing in C. elegans and is regulated by the lin-4 RNA. Cell *88*, 637-646.

Napoli, C., Lemieux, C., and Jorgensen, R. (1990). Introduction of a Chimeric Chalcone Synthase Gene into Petunia Results in Reversible Co-Suppression of Homologous Genes in trans. Plant Cell *2*, 279-289.

Nellen, W., and Lichtenstein, C. (1993). What makes an mRNA anti-sense-itive? Trends Biochem Sci *18*, 419-423.

Palatnik, J.F., Allen, E., Wu, X., Schommer, C., Schwab, R., Carrington, J.C., and Weigel, D. (2003). Control of leaf morphogenesis by microRNAs. Nature *425*, 257-263.

Palatnik, J.F., Wollmann, H., Schommer, C., Schwab, R., Boisbouvier, J., Rodriguez, R., Warthmann, N., Allen, E., Dezulian, T., Huson, D., *et al.* (2007). Sequence and expression differences underlie functional specialization of arabidopsis microRNAs miR159 and miR319. Dev Cell *13*, 115-125.

Park, M.Y., Wu, G., Gonzalez-Sulser, A., Vaucheret, H., and Poethig, R.S. (2005). Nuclear processing and export of microRNAs in Arabidopsis. Proc Natl Acad Sci USA *102*, 3691-3696.

Park, W., Li, J., Song, R., Messing, J., and Chen, X. (2002). CARPEL FACTORY, a Dicer homolog, and HEN1, a novel protein, act in microRNA metabolism in Arabidopsis thaliana. Curr Biol *12*, 1484-1495.

Pasquinelli, A.E., Reinhart, B.J., Slack, F., Martindale, M.Q., Kuroda, M.I., Maller, B., Hayward, D.C., Ball, E.E., Degnan, B., Muller, P., *et al.* (2000). Conservation of the sequence and temporal expression of let-7 heterochronic regulatory RNA. Nature *408*, 86-89.

References

Peragine, A., Yoshikawa, M., Wu, G., Albrecht, H.L., and Poethig, R.S. (2004). SGS3 and SGS2/SDE1/RDR6 are required for juvenile development and the production of trans-acting siRNAs in Arabidopsis. Genes Dev *18*, 2368-2379.

Pouch-Pelissier, M.N., Pelissier, T., Elmayan, T., Vaucheret, H., Boko, D., Jantsch, M.F., and Deragon, J.M. (2008). SINE RNA induces severe developmental defects in Arabidopsis thaliana and interacts with HYL1 (DRB1), a key member of the DCL1 complex. PLoS genetics *4*, e1000096.

Prigge, M.J., and Wagner, D.R. (2001). The arabidopsis serrate gene encodes a zinc-finger protein required for normal shoot development. Plant Cell *13*, 1263-1279.

Qi, Y., Denli, A.M., and Hannon, G.J. (2005). Biochemical specialization within Arabidopsis RNA silencing pathways. Mol Cell *19*, 421-428.

Rajagopalan, R., Vaucheret, H., Trejo, J., and Bartel, D.P. (2006). A diverse and evolutionarily fluid set of microRNAs in Arabidopsis thaliana. Genes Dev *20*, 3407-3425.

Ramachandran, V., and Chen, X. (2008). Degradation of microRNAs by a family of exoribonucleases in Arabidopsis. Science *321*, 1490-1492.

Reinhart, B.J., Slack, F.J., Basson, M., Pasquinelli, A.E., Bettinger, J.C., Rougvie, A.E., Horvitz, H.R., and Ruvkun, G. (2000). The 21-nucleotide let-7 RNA regulates developmental timing in Caenorhabditis elegans. Nature *403*, 901-906.

Rodriguez-Trelles, F., Tarrio, R., and Ayala, F.J. (2005). Is ectopic expression caused by deregulatory mutations or due to gene-regulation leaks with evolutionary potential? Bioessays *27*, 592-601.

Ruiz-Ferrer, V., and Voinnet, O. (2009). Roles of plant small RNAs in biotic stress responses. Annu Rev Plant Biol *60*, 485-510.

Schauer, S.E., Jacobsen, S.E., Meinke, D.W., and Ray, A. (2002). DICER-LIKE1: blind men and elephants in Arabidopsis development. Trends Plant Sci *7*, 487-491.

Schneider, T.D., and Stephens, R.M. (1990). Sequence logos: a new way to display consensus sequences. Nucleic Acids Res *18*, 6097-6100.

Schommer, C., Palatnik, J.F., Aggarwal, P., Chetelat, A., Cubas, P., Farmer, E.E., Nath, U., and Weigel, D. (2008). Control of jasmonate biosynthesis and senescence by miR319 targets. PLoS Biol *6*, e230.

References

Schwab, R., Ossowski, S., Riester, M., Warthmann, N., and Weigel, D. (2006). Highly specific gene silencing by artificial microRNAs in Arabidopsis. Plant Cell *18*, 1121-1133.

Schwab, R., Palatnik, J.F., Riester, M., Schommer, C., Schmid, M., and Weigel, D. (2005). Specific effects of microRNAs on the plant transcriptome. Dev Cell *8*, 517-527.

Seydoux, G., and Braun, R.E. (2006). Pathway to totipotency: lessons from germ cells. Cell *127*, 891-904.

Sieber, P., Wellmer, F., Gheyselinck, J., Riechmann, J.L., and Meyerowitz, E.M. (2007). Redundancy and specialization among plant microRNAs: role of the MIR164 family in developmental robustness. Development *134*, 1051-1060.

Song, L., Han, M.H., Lesicka, J., and Fedoroff, N. (2007). Arabidopsis primary microRNA processing proteins HYL1 and DCL1 define a nuclear body distinct from the Cajal body. Proc Natl Acad Sci USA.

Souret, F.F., Kastenmayer, J.P., and Green, P.J. (2004). AtXRN4 degrades mRNA in Arabidopsis and its substrates include selected miRNA targets. Mol Cell *15*, 173-183.

Sun, G., Yan, J., Noltner, K., Feng, J., Li, H., Sarkis, D.A., Sommer, S.S., and Rossi, J.J. (2009). SNPs in human miRNA genes affect biogenesis and function. Rna *15*, 1640-1651.

Takeda, A., Iwasaki, S., Watanabe, T., Utsumi, M., and Watanabe, Y. (2008). The mechanism selecting the guide strand from small RNA duplexes is different among argonaute proteins. Plant Cell Physiol *49*, 493-500.

Vaucheret, H. (2008). Plant ARGONAUTES. Trends Plant Sci *13*, 350-358.

Vaucheret, H., Vazquez, F., Crete, P., and Bartel, D.P. (2004). The action of ARGONAUTE1 in the miRNA pathway and its regulation by the miRNA pathway are crucial for plant development. Genes Dev *18*, 1187-1197.

Vazquez, F., Blevins, T., Ailhas, J., Boller, T., and Meins, F., Jr. (2008). Evolution of Arabidopsis MIR genes generates novel microRNA classes. Nucleic Acids Res *36*, 6429-6438.

Vazquez, F., Gasciolli, V., Crété, P., and Vaucheret, H. (2004a). The nuclear dsRNA binding protein HYL1 is required for microRNA accumulation and plant development, but not posttranscriptional transgene silencing. Curr Biol *14*, 346-351.

References

Vazquez, F., Vaucheret, H., Rajagopalan, R., Lepers, C., Gasciolli, V., Mallory, A.C., Hilbert, J.L., Bartel, D.P., and Crete, P. (2004b). Endogenous trans-acting siRNAs regulate the accumulation of Arabidopsis mRNAs. Mol Cell 16, 69-79.

Voinnet, O. (2009). Origin, biogenesis, and activity of plant microRNAs. Cell 136, 669-687.

Wang, J.W., Czech, B., and Weigel, D. (2009). miR156-regulated SPL transcription factors define an endogenous flowering pathway in Arabidopsis thaliana. Cell 138, 738-749.

Warthmann, N., Das, S., Lanz, C., and Weigel, D. (2008). Comparative analysis of the MIR319a microRNA locus in Arabidopsis and related Brassicaceae. Mol Biol Evol 25, 892-902.

Wells, S.E., Hillner, P.E., Vale, R.D., and Sachs, A.B. (1998). Circularization of mRNA by eukaryotic translation initiation factors. Mol Cell 2, 135-140.

Wightman, B., Ha, I., and Ruvkun, G. (1993). Posttranscriptional regulation of the heterochronic gene lin-14 by lin-4 mediates temporal pattern formation in C. elegans. Cell 75, 855-862.

Winter, J., Jung, S., Keller, S., Gregory, R.I., and Diederichs, S. (2009). Many roads to maturity: microRNA biogenesis pathways and their regulation. Nat Cell Biol 11, 228-234.

Wu, G., Park, M.Y., Conway, S.R., Wang, J.W., Weigel, D., and Poethig, R.S. (2009). The sequential action of miR156 and miR172 regulates developmental timing in Arabidopsis. Cell 138, 750-759.

Xie, Z., Allen, E., Fahlgren, N., Calamar, A., Givan, S.A., and Carrington, J.C. (2005). Expression of Arabidopsis MIRNA genes. Plant Physiol 138, 2145-2154.

Xie, Z., Johansen, L.K., Gustafson, A.M., Kasschau, K.D., Lellis, A.D., Zilberman, D., Jacobsen, S.E., and Carrington, J.C. (2004). Genetic and functional diversification of small RNA pathways in plants. PLoS Biol 2, E104.

Xie, Z., Kasschau, K.D., and Carrington, J.C. (2003). Negative feedback regulation of Dicer-Like1 in Arabidopsis by microRNA-guided mRNA degradation. Curr Biol 13, 784-789.

Yanai, I., Korbel, J.O., Boue, S., McWeeney, S.K., Bork, P., and Lercher, M.J. (2006). Similar gene expression profiles do not imply similar tissue functions. Trends Genet 22, 132-138.

Yang, L., Liu, Z., Lu, F., Dong, A., and Huang, H. (2006a). SERRATE is a novel nuclear regulator in primary microRNA processing in Arabidopsis. Plant J.

References

Yang, Z., Ebright, Y.W., Yu, B., and Chen, X. (2006b). HEN1 recognizes 21-24 nt small RNA duplexes and deposits a methyl group onto the 2' OH of the 3' terminal nucleotide. Nucleic Acids Res *34*, 667-675.

Yu, B., Bi, L., Zheng, B., Ji, L., Chevalier, D., Agarwal, M., Ramachandran, V., Li, W., Lagrange, T., Walker, J.C., *et al.* (2008). The FHA domain proteins DAWDLE in Arabidopsis and SNIP1 in humans act in small RNA biogenesis. Proc Natl Acad Sci USA *105*, 10073-10078.

Zamore, P.D., Tuschl, T., Sharp, P.A., and Bartel, D.P. (2000). RNAi: double-stranded RNA directs the ATP-dependent cleavage of mRNA at 21 to 23 nucleotide intervals. Cell *101*, 25-33.

Zhao, L., Kim, Y., Dinh, T.T., and Chen, X. (2007). miR172 regulates stem cell fate and defines the inner boundary of APETALA3 and PISTILLATA expression domain in Arabidopsis floral meristems. Plant J *51*, 840-849.

Zuker, M. (2003). Mfold web server for nucleic acid folding and hybridization prediction. Nucleic Acids Res *31*, 3406-3415.

MATERIALS AND METHODS

Standard techniques and buffers

The preparation of standard buffers and media, and standard molecular techniques, were performed according to (Sambrook et al., 1989). All chemicals were purchased from Sigma (Munich, Germany), Bio-Rad (Munich, Germany), Roth (Karlsruhe, Germany), Merck (Darmstadt, Germany) and Roche (Mannheim, Germany). Restriction endonucleases were purchased from Fermentas (Burlington, Canada) and New England Biolabs (Ipswich, MA, USA). DNA Polymerases were purchased from Fermentas (Pfu, Taq) and Finnzyme (Espoo, Finland; Phusion). Oligonucleotides were ordered from MWG (Ebersberg, Germany). LNA oligonucleotides were ordered from Exiqon (Vedbaek, Denmark).

DNA extraction

Plant material was harvested and grinded. DNA extraction was carried out by first adding 400µl DNA extraction buffer (0.2M Tris/HCL pH7.5, 0.25M NaCL, 25mM EDTA, 0.5% (w/v) SDS), after mixing, the eppendorf-tube was centrifuged at full speed for 3min. 300µl were transferred into a new tube with 300µl Isopropanol. After 2min incubation at room temperature, the DNA was pelleted by full speed centrifugation for 10min. After one washing step with 75% ethanol, the extracted DNA was resuspended in 100µl water.

Plant transformation

Transgenic plants were generated by *Agrobacterium tumefaciens* mediated transformation using the strains ASE or GV3101. Plants were transformed by floral dip (Clough and Bent, 1998). At least 15 T1 plants were analyzed for each construct.

Plant growth

Plants were generally grown in long day (16 hours light, 8 hours dark) conditions at 23°C with 65% humidity. *Arabidopsis thaliana* plants of the accession Col-0 were used as wild-type plants. Before sowing, seeds were sterilized by incubation at -20°C for at least two nights or by ethanol sterilization. Sterilized seeds were kept for three to five days at 4°C in 0.1% agarose for stratification. For selection of BASTA resistant plants, a 1:1000 dilution was applied to the initial water for soaking the soil.

Materials and Methods

Molecular Cloning

Escherichia coli strain DH5α (Life Technologies) was used for plasmid amplification. Plasmid DNA was extracted by alkaline lysis or using the Wizard® Plus SV Minipreps Kit (Promega, Mannheim, Germany). DNA from agarose gels was isolated with the Wizard® SV Gel and PCR clean-up system (Promega) or by squeezing the liquid from the agarose gel and subsequent DNA precipitation with SureClean (Bioline).

Antibiotics were used in the following final concentrations for bacterial growth:
- 100 µg/ml Ampicillin
- 25   µg/ml Chloramphenicol
- 50   µg/ml Kanamycin
- 100 µg/ml Spectinomycin

Mutant constructs were generated from the plasmid HW042 (Gateway entry vector with pri-miR172a) by DpnI mediated site directed mutagenesis and PCR amplification (for plasmid sequence see below). Mutant precursors were recombined into pGREEN-IIS destination vectors, with the CaMV 35S promoter in front of a modified Gateway recombination cassette. pGREEN-IIS, a derivative of pGREEN-II (Hellens et al., 2000), confers resistance to spectinomycin in bacteria. Sequences of mutant constructs are available on request.

Sequence of plasmid HW042

Pri-miR172a sequence in bold:

```
GCCTACATACCTCGCTCTGCTAATCCTGTTACCAGTGGCTGCTGCCAGTGGCGATAAGTCGTGTCTTAC
CGGGTTGGACTCAAGACGATAGTTACCGGATAAGGCGCAGCGGTCGGGCTGAACGGGGGGTTCGTGC
ACACAGCCCAGCTTGGAGCGAACGACCTACACCGAACTGAGATACCTACAGCGTGAGCTATGAGAAAG
CGCCACGCTTCCCGAAGGGAGAAAGGCGGACAGGTATCCGGTAAGCGGCAGGGTCGGAACAGGAGA
GCGCACGAGGGAGCTTCCAGGGGGAAACGCCTGGTATCTTTATAGTCCTGTCGGGTTTCGCCACCTCT
GACTTGAGCGTCGATTTTGTGATGCTCGTCAGGGGGGCGGAGCCTATGGAAAAACGCCAGCAACGCG
GCCTTTTTACGGTTCCTGGCCTTTTGCTGGCCTTTTGCTCACATGTTCTTTCCTGCGTTATCCCCTGATT
CTGTGGATAACCGTATTACCGCTAGCATGGATCTCGGGGACGTCTAACTACTAAGCGAGAGTAGGGAA
CTGCCAGGCATCAAATAAAACGAAAGGCTCAGTCGGAAGACTGGGCCTTTCGTTTTATCTGTTGTTTGT
CGGTGAACGCTCTCCTGAGTAGGACAAATCCGCCGGGAGCGGATTTGAACGTTGTGAAGCAACGGCCC
GGAGGGTGGCGGGCAGGACGCCCGCCATAAACTGCCAGGCATCAAACTAAGCAGAAGGCCATCCTGA
CGGATGGCCTTTTTGCGTTTCTACAAACTCTTCCTGTTAGTTAGTTACTTAAGCTCGGGCCCCAAATAAT
GATTTTATTTTGACTGATAGTGACCTGTTCGTTGCAACAAATTGATAAGCAATGCTTTTTTATAATGCCAA
CTTTGTACAAAAAAGCAGGCTTTCGAATTCCAAGCTTGCCCCGACGGTATCGATAAGCTTGATATCGAAT
```

Materials and Methods

TCCTGCAGCCC**AAAAATGGAAGACTAATTTCCGGAGCCACGGT**cgttgttggctgctgtggcatcatcaagattcac
atctgttgatggacggtggtgattcactctccacaaagttctctatgaaaatg**AGAATCTTGATGATGCTGCAT**cggcaatcaacg
ACTATTCTTTCCCTCTCTCTCTCTCCCTCTGTATAGATTATTTGGATTCCATCCAGATCTTCTTCAGG
TAGGTTTGTTTCTACTTGAAGTTTTTTTTTTTCACCTTTATGTTAACATATCCTCCCGTTTATTTTTATTTG
TTATACATAAAGATCTGACAAAGAACTTTTGTGGGTATCTTGTTTCATGTGATAACATTGAGCATTTGA
TCTCAGGTTTTTGGCAGTCTTTATCAAGACATTAATAGATCCACAAGCGGGGGATGGGCGGATCCGAA
TTCCTATCTAGACCCAGCTTTCTTGTACAAAGTTGGCATTATAAGAAAGCATTGCTTATCAATTTGTTGCA
ACGAACAGGTCACTATCAGTCAAAATAAAATCATTATTTGCCATCCAGCTGCAGCTCTGGCCCGTGTCTC
AAAATCTCTGATGTTACATTGCACAAGATAAAAATATATCATCATGAACAATAAAACTGTCTGCTTACATA
AACAGTAATACAAGGGGTGTTATGAGCCATATTCAACGGGAAACGTCGAGGCCGCGATTAAATTCCAAC
ATGGATGCTGATTTATATGGGTATAAATGGGCTCGCGATAATGTCGGGCAATCAGGTGCGACAATCTAT
CGCTTGTATGGGAAGCCCGATGCGCCAGAGTTGTTTCTGAAACATGGCAAAGGTAGCGTTGCCAATGA
TGTTACAGATGAGATGGTCAGACTAAACTGGCTGACGGAATTTATGCCTCTTCCGACCATCAAGCATTTT
ATCCGTACTCCTGATGATGCATGGTTACTCACCACTGCGATCCCCGGAAAAACAGCATTCCAGGTATTA
GAAGAATATCCTGATTCAGGTGAAAATATTGTTGATGCGCTGGCAGTGTCCCTGCGCCGGTTGCATTCG
ATTCCTGTTTGTAATTGTCCTTTTAACAGCGATCGCGTATTTCGTCTCGCTCAGGCGCAATCACGAATGA
ATAACGGTTTGGTTGATGCGAGTGATTTTGATGACGAGCGTAATGGCTGGCCTGTTGAACAAGTCTGGA
AAGAAATGCATAAACTTTTGCCATTCTCACCGGATTCAGTCGTCACTCATGGTGATTTCTCACTTGATAA
CCTTATTTTTGACGAGGGGAAATTAATAGGTTGTATTGATGTTGGACGAGTCGGAATCGCAGACCGATA
CCAGGATCTTGCCATCCTATGGAACTGCCTCGGTGAGTTTTCTCCTTCATTACAGAAACGGCTTTTTCAA
AAATATGGTATTGATAATCCTGATATGAATAAATTGCAGTTTCATTTGATGCTCGATGAGTTTTTCTAATC
AGAATTGGTTAATTGGTTGTAACATTATTCAGATTGGGCCCCGTTCCACTGAGCGTCAGACCCGGTAGA
AAAGATCAAAGGATCTTCTTGAGATCCTTTTTTTCTGCGCGTAATCTGCTGCTTGCAAACAAAAAAACCA
CCGCTACCAGCGGTGGTTTGTTTGCCGGATCAAGAGCTACCAACTCTTTTTCCGAAGGTAACTGGCTTC
AGCAGAGCGCAGATACCAAATACTGTTCTTCTAGTGTAGCCGTAGTTAGGCCACCACTTCAAGAACTCT
GTAGCACC

RNA analysis

Total RNA was isolated from inflorescences of at least 12 pooled T1 plants using TRIZOL® reagent (Invitrogen, Carlsbad, CA). Small RNAs were detected by small RNA blots with end-labeled oligonucleotide probes (DNA for U6 rRNA detection, and LNA [Exiqon, Vedbaek, Denmark] for miR172a and amiR-CH42 detection). Cleavage sites in pri-miR172a were mapped by 5' RACE (rapid amplification of cloned ends) like described in (Llave et al., 2002) (see Figure S2e). Primers used for first PCR and second PCR (nested):

Forward: Primers of 5' adapter GeneRacer™ Kit (Invitrogen)
Reverse: G-20525   TGGTGTGTGCGCAATGAAACTGATGC (1. PCR)
          G-01209   CTGAAGAAGATCTGGATGGAATCC (2. PCR)

Small RNA blots

4 µg total RNA were loaded on 17 % polyacrylamide gel with 7 M urea. 1.5 volume loading dye with 2 volumes formamide were added to the samples, which were heated to 95 °C for 4 minutes then cooled down on ice. A pre-run at 150 volts for 1 hour in 0.5 x TBE was done, before the actual run was performed (Mini-Protean®3 Electrophoresis Cell (Bio-Rad)). As loading control, gels were stained with ethidium bromide to visualize the RNA and documented. After a brief de-stain of the gel in 0.5 x TBE, the RNA was transferred to Nytran supercharge membrane (Schleicher and Schuell) with a Semi-dry blotting system (Bio-Rad). After transfer, RNA was cross-linked (UV stratalinker). Antisense LNA oligonucleotides were radioactively end-labeled with the OptiKinaseTM (USB), purified with MicroBio-Spin®6 Chromatography Columns (Bio-Rad) and used as probes. Probes were mixed for hybridization. Pre-hybridization in PerfectHybTMPlus (Sigma) was carried out for at least 1 hour at 38 °C before adding the probe. Hybridization was carried out over night, followed by washes with 2 x SSC/0.2 % SDS. Exposure to BiomaxMS films (Kodak) was carried out at -80°C.

MiRNA secondary structure analysis

All miRNA structures were folded with *mfold* 3.2 (Zuker, 2003). For all *A. thaliana* miRNAs conserved in *Populus trichocarpa* (Fahlgren et al., 2007), with the exception o the miR159 and miR319 families, the annotated miRNA stem sequence were extended and folded with RNAfold (ViennaRNA 1.8.3) (Hofacker et al., 1994). A position weight matrix analysis was performed separately for 5' and 3' arms. A value was assigned based on the size of the unpaired region that a base participated in, with 0 for paired bases, 1 for unpaired regions of size 1, 2 for unpaired regions of size 2, etc.

Visualisation of a hypothetical common sequence motif

To visualise the sequence analyses in chapter II I used the program Weblogo (http://weblogo.berkeley.edu/). It shows a graphical representation of nucleic acid multiple sequence alignments. Each logo consists of stacks of symbols, one stack for each position in the sequence. The overall height of the symbols within the stack indicates the relative frequency of each nucleic acid at that position.

Sliding window method

To analyse average energy profiles of miRNA foldbacks I used the sliding window method with a window size of 21 nucleotides (the length of miRNAs).

SUPPLEMENTARY MATERIAL

Inventory of Supplemental Information

Supplementary Table S1
Key for diagrams and photos of miR172a foldback mutants in Supplementary Figure 3 (related to Supplemental Figure 2b, 2c, 2d, 3).

Supplementary Figure S1a
Secondary structure of the full-length transcript of the cloned MIR172a

Supplementary Figure S1b
MiR172a foldbacks generated at different temperatures with *mfold*3.2

Supplementary Figure S1c
miR172a overexpressor mutants isolated from EMS screen.

Supplementary Figure S2a
Flowering time overview of all transformed plant lines (related to Supplemental Figure 1c, 2b, 2c, 2d, 3)

Supplementary Figure S2b
Mutations in the distal miR172a foldback region

Supplementary Figure S2c
Mutations in the miR172a duplex region.

Supplementary Figure S2d
Mutations in the proximal miR172a foldback region

Supplementary Figure S2e
5' RACE diagram of the cloned MIR172a.

Supplementary Material

Supplementary Figure S3

Foldbacks and phenotypes of individual mutants (Supplemental Figure 2b, 2c, 2d).

Supplementary Figure S4

Foldback structure of *Arabidopsis lyrata* pri-miR172a

Supplementary Material

Supplementary Table 1:

Key for diagrams and photos of miR172a foldback mutants in Suppl. Fig. 3.

Panel	Mutant	Region	Panel	Mutant	Region
a	wild type	-	ae	26	Proximal
b	-4nt	Proximal	af	27	Proximal
c	stem1	Proximal	ag	28	Proximal
d	stem2	Proximal	ah	29	Proximal
e	stemcore	Proximal	ai	30	Proximal
f	1	miRNA Duplex	aj	31	miRNA Duplex
g	2	miRNA Duplex	ak	32	miRNA Duplex
h	3	miRNA Duplex	al	33	miRNA Duplex
i	4	miRNA Duplex	am	34	miRNA Duplex
j	5	miRNA Duplex	an	35	miRNA Duplex
k	6	miRNA Duplex	ao	36	miRNA Duplex
l	7	miRNA Duplex	ap	38	miRNA Duplex
m	8	Distal	aq	39	miRNA Duplex
n	9	Proximal	ar	40	miRNA Duplex
o	10	Proximal	as	41	miRNA Duplex
p	11	Duplex/Proximal	at	42	miRNA Distal
q	12	Duplex/Distal	au	43	miRNA Duplex
r	13	Duplex/Distal	av	48	miRNA Duplex
s	14	Distal	aw	49	miRNA Duplex
t	15	Distal	ax	50	miRNA Duplex
u	16	miRNA Duplex	ay	52	Duplex/Proximal
v	17	miRNA Duplex	az	53	miRNA Duplex
w	18	Proximal	ba	54	Proximal
x	19	Proximal	bb	55	Proximal

Supplementary Material

y	20	Proximal	bc	57	Distal
z	21	Proximal	bd	60	Proximal
aa	22	Proximal	be	61	Proximal
ab	23	Proximal	bf	62	Proximal
ac	24	Proximal	bg	63	Distal
ad	25	Proximal			

Supplementary Material

Supplementary Figure 1a

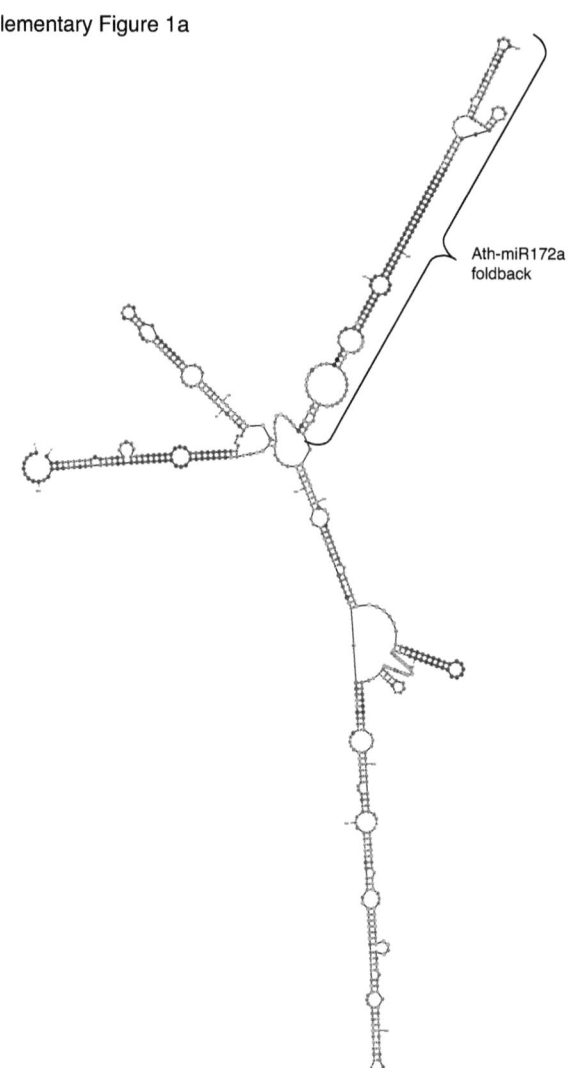

Supplementary Figure 1b

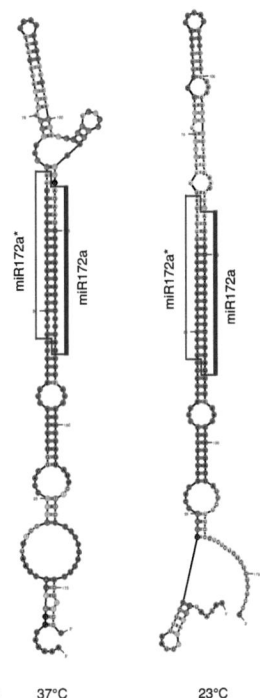

folded with *mfold* 3.2 at: 37°C   23°C

Supplementary Material

Supplementary Figure 1c

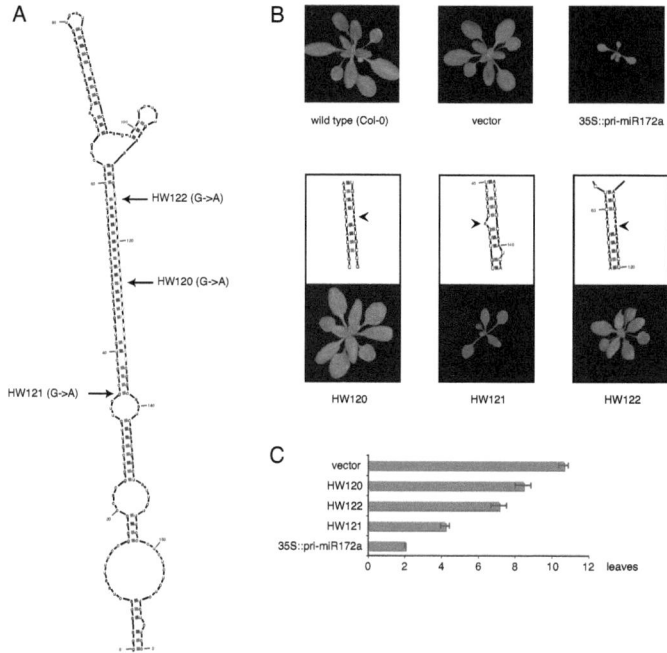

Supplementary Material

Supplementary Figure 2a

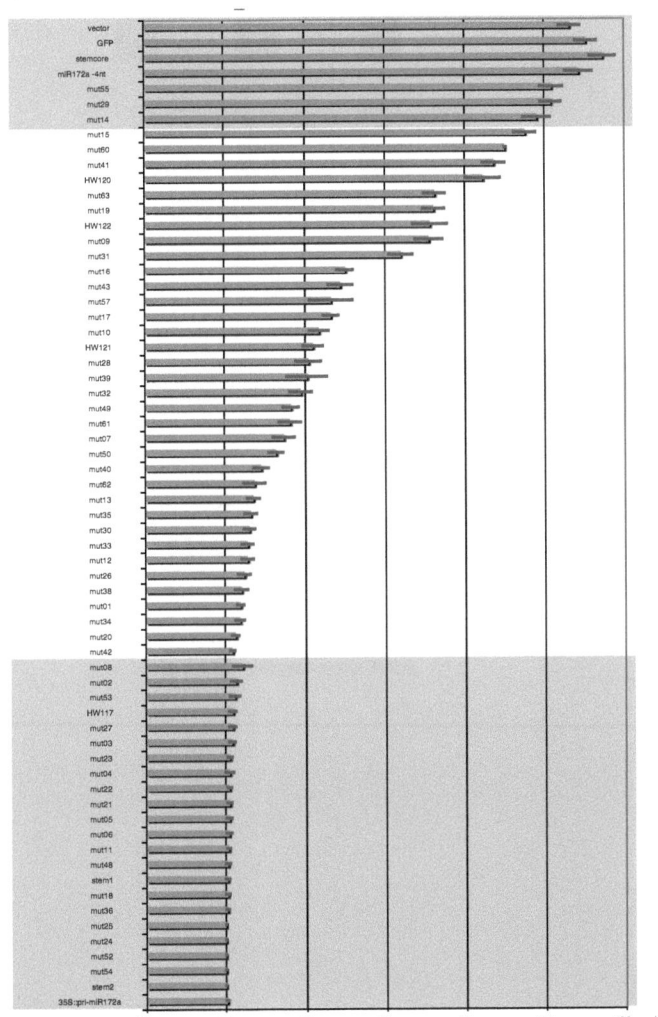

Supplementary Material

Supplementary Figure 2b

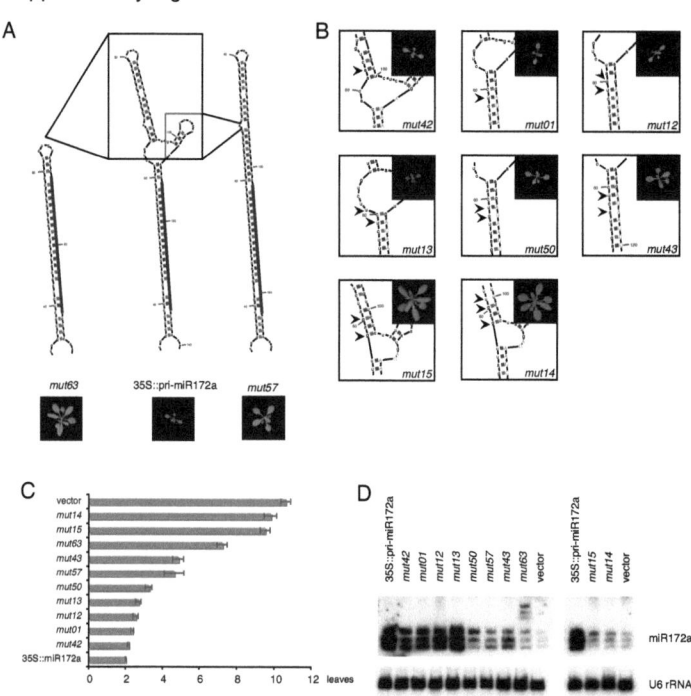

Supplementary Material

Supplementary Figure 2c

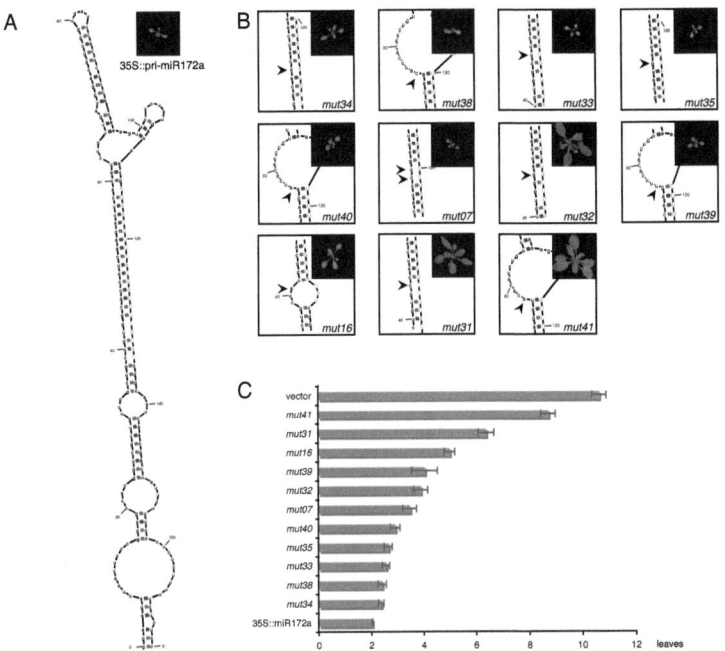

Supplementary Material

Supplementary Figure 2d

A

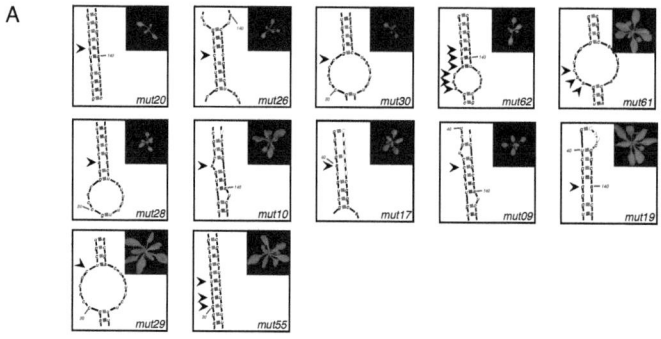

B

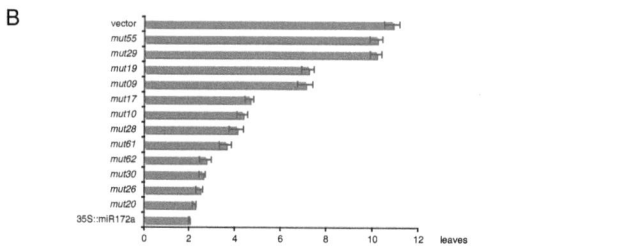

C

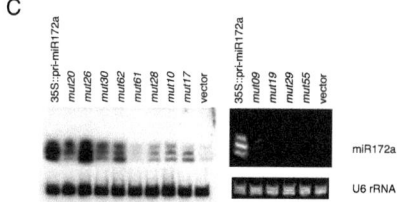

Supplementary Figure 2e

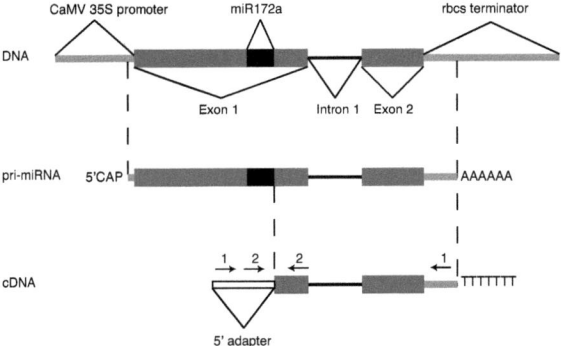

Supplementary Figure 3a   **miR172a foldback wild-type**

flowering time (rosette leaves):
2.0 (±0.02 SEM; n = 64)

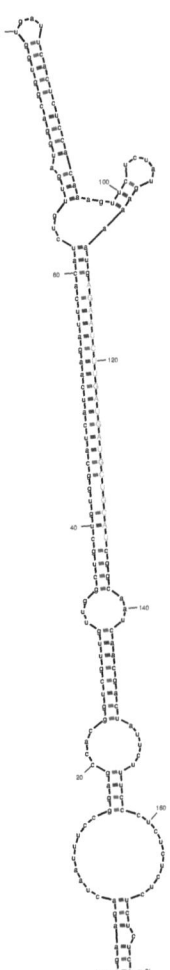

Supplementary Material

Supplementary Figure 3b   **mutant -4nt**

flowering time (rosette leaves):
10.9 (±0.34 SEM; n = 15)

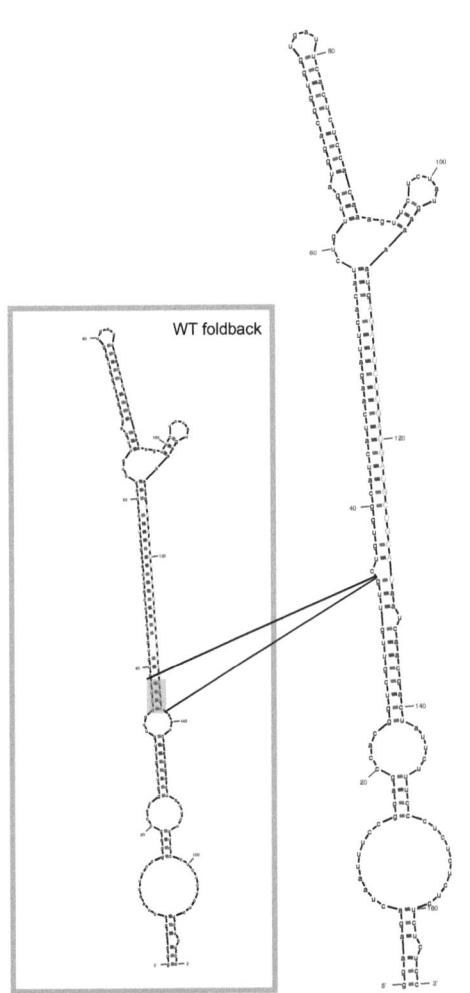

WT foldback

Supplementary Figure 3c — mutant stem1

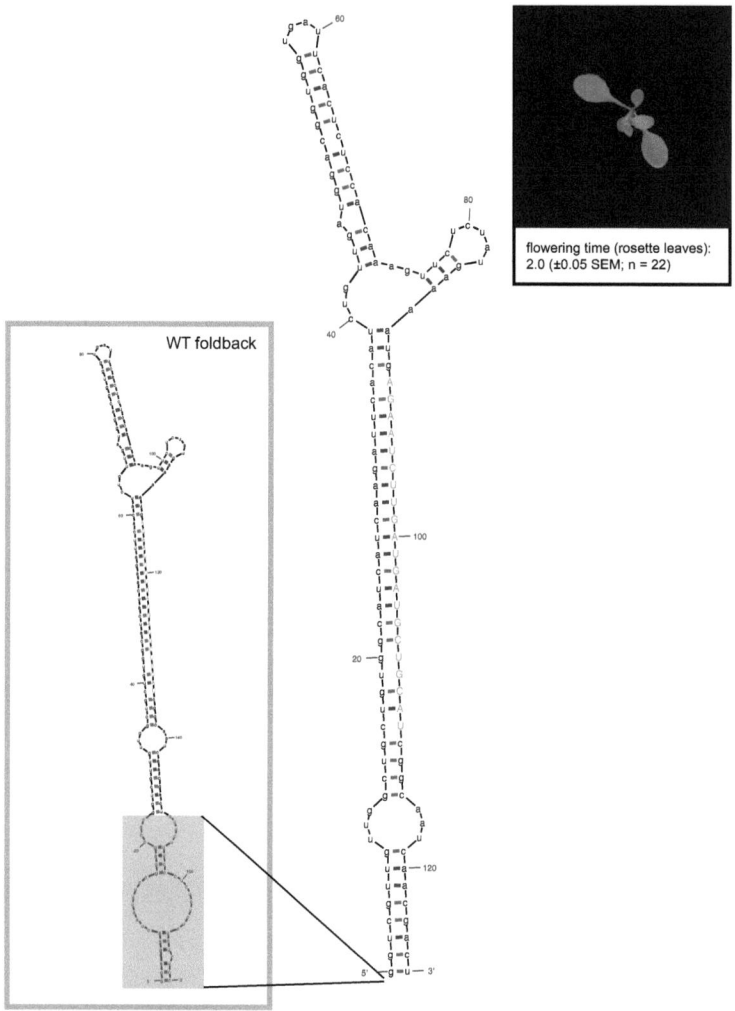

Supplementary Figure 3d

mutant stem2

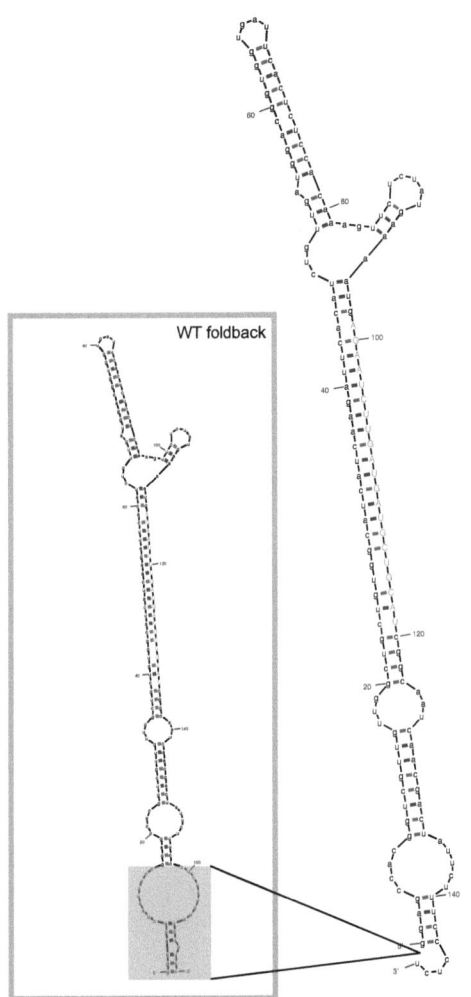

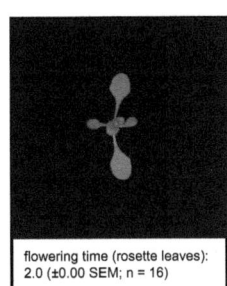

flowering time (rosette leaves):
2.0 (±0.00 SEM; n = 16)

WT foldback

Supplementary Figure 3e — mutant stemcore

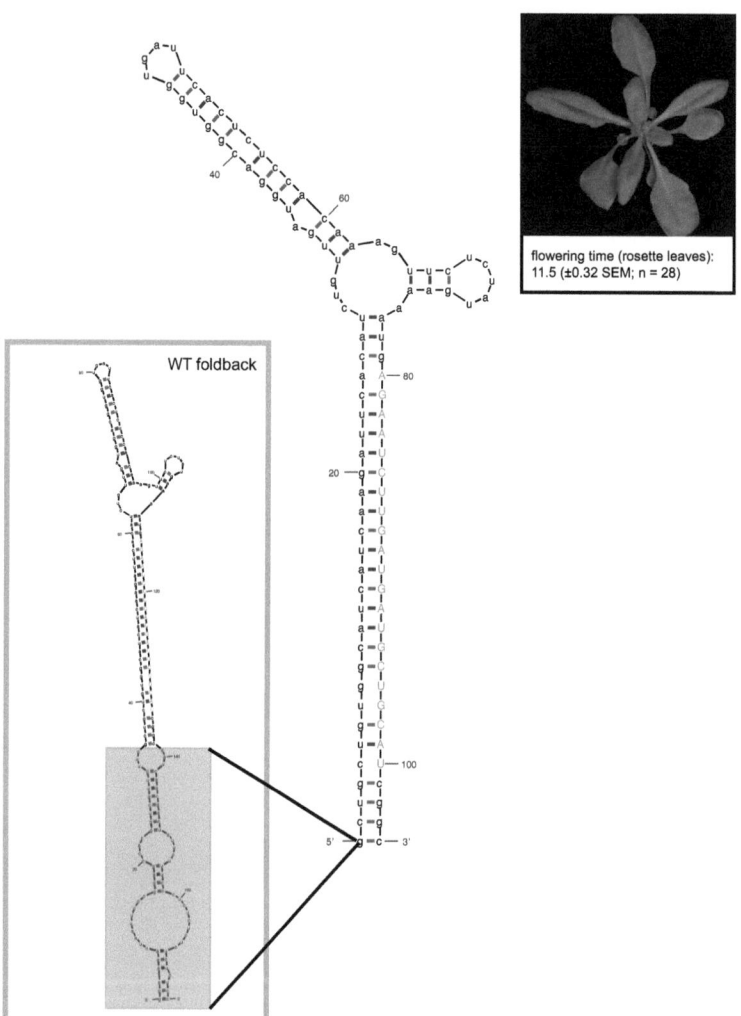

flowering time (rosette leaves):
11.5 (±0.32 SEM; n = 28)

WT foldback

Supplementary Figure 3f mutant 01

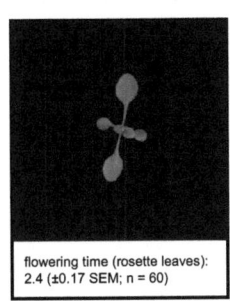

flowering time (rosette leaves):
2.4 (±0.17 SEM; n = 60)

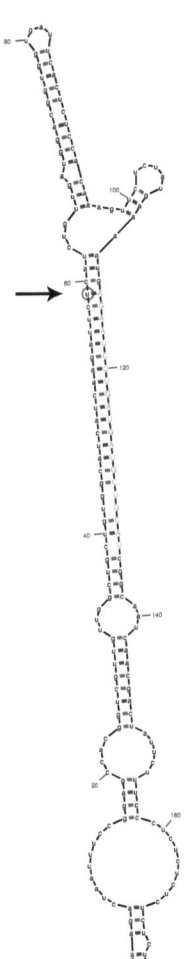

Supplementary Material

Supplementary Figure 3g — mutant 02

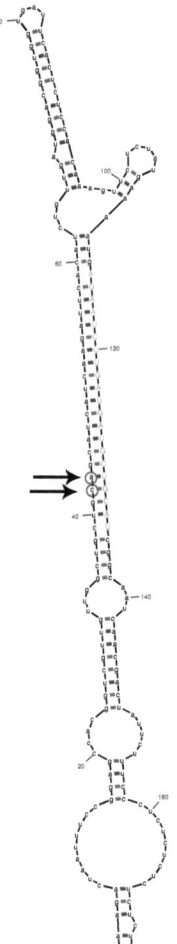
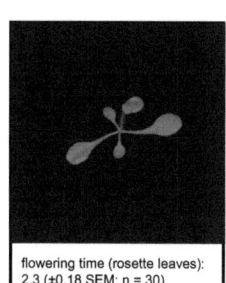

flowering time (rosette leaves):
2.3 (±0.18 SEM; n = 30)

Supplementary Material

Supplementary Figure 3h

mutant 03

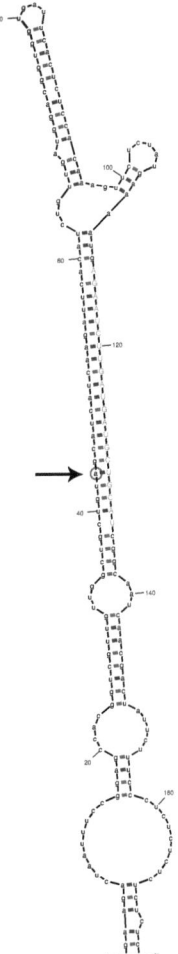
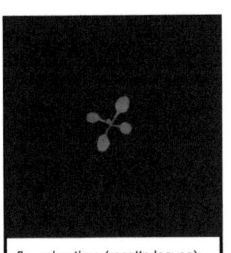

flowering time (rosette leaves):
2.1 (±0.13 SEM; n = 42)

Supplementary Figure 3i   mutant 04

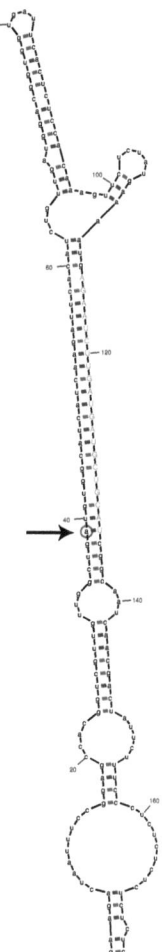

flowering time (rosette leaves):
2.1 (±0.2 SEM; n = 47)

Supplementary Material

Supplementary Figure 3j   mutant 05

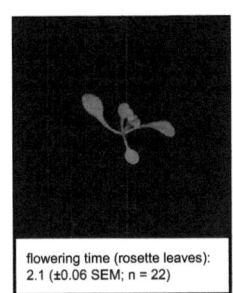

flowering time (rosette leaves):
2.1 (±0.06 SEM; n = 22)

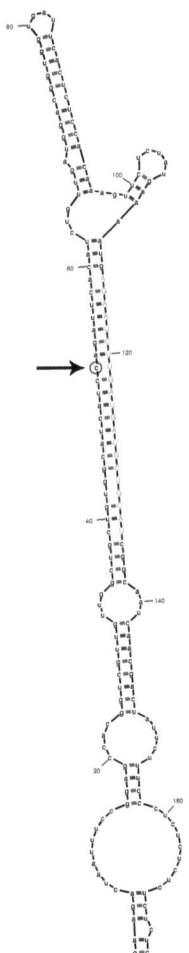

Supplementary Figure 3k — mutant 06

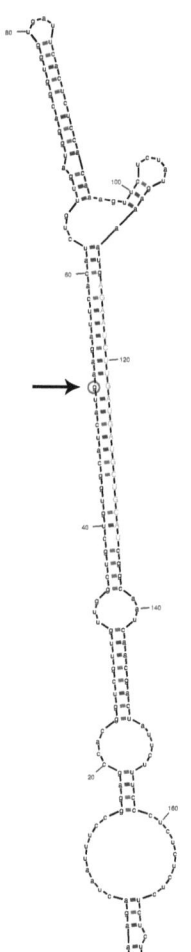
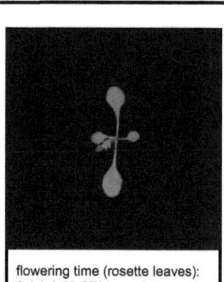

flowering time (rosette leaves):
2.1 (±0.06 SEM; n = 22)

Supplementary Figure 3l   **mutant 07**

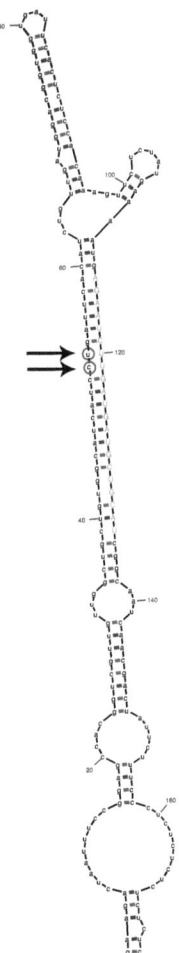

flowering time (rosette leaves):
3.5 (±0.26 SEM; n = 41)

Supplementary Material

Supplementary Figure 3m   **mutant 08**

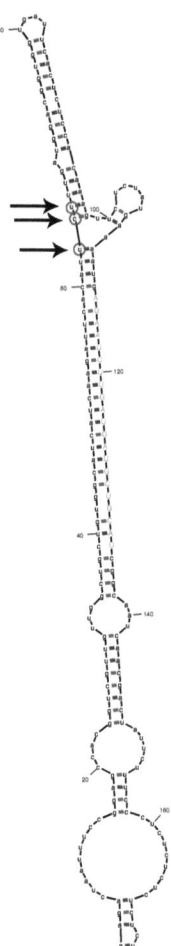
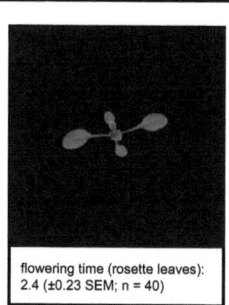

flowering time (rosette leaves):
2.4 (±0.23 SEM; n = 40)

Supplementary Figure 3n   mutant 09

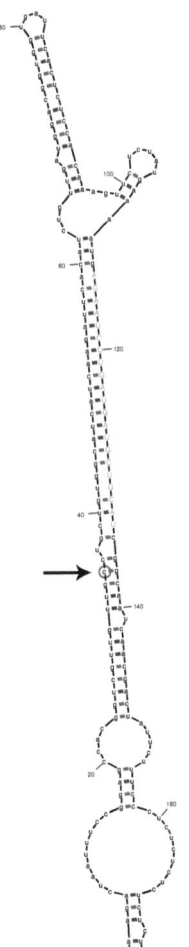

flowering time (rosette leaves):
7.1 (±0.34 SEM; n = 63)

Supplementary Figure 3o   mutant 10

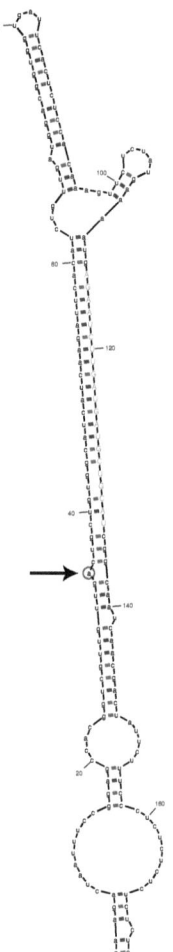

flowering time (rosette leaves):
4.3 (±0.24 SEM; n = 41)

Supplementary Figure 3p — mutant 11

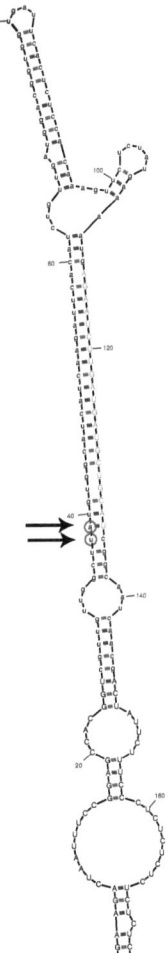
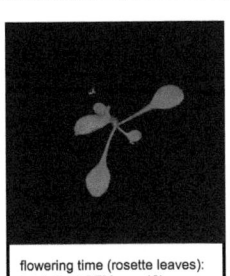

flowering time (rosette leaves):
2.1 (±0.04 SEM; n = 42)

Supplementary Material

Supplementary Figure 3q — mutant 12

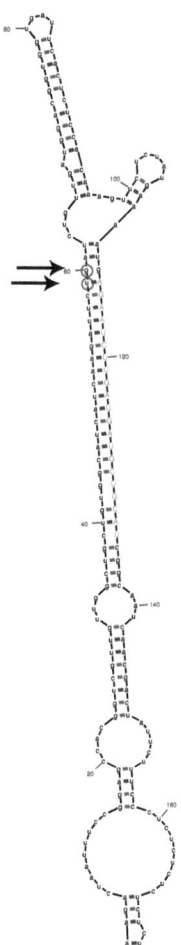
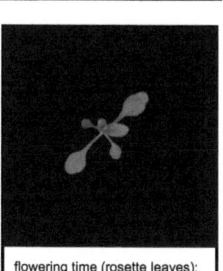

flowering time (rosette leaves):
2.6 (±0.14 SEM; n = 41)

Supplementary Figure 3r   **mutant 13**

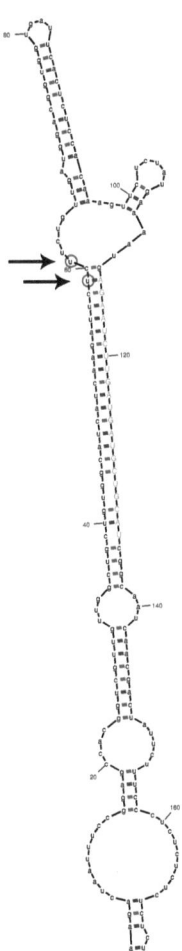

flowering time (rosette leaves):
2.7 (±0.15 SEM; n = 41)

Supplementary Material

Supplementary Figure 3s   **mutant 14**

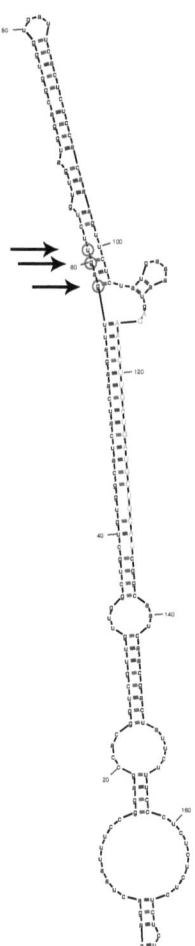

flowering time (rosette leaves):
9.1 (±0.33 SEM; n = 42)

Supplementary Figure 3t
mutant 15

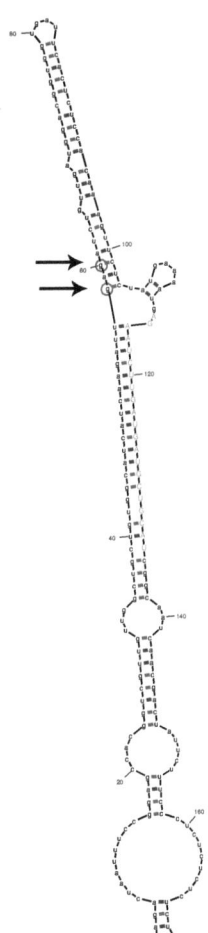

flowering time (rosette leaves):
9.5 (±0.27 SEM; n = 41)

Supplementary Figure 3u — mutant 16

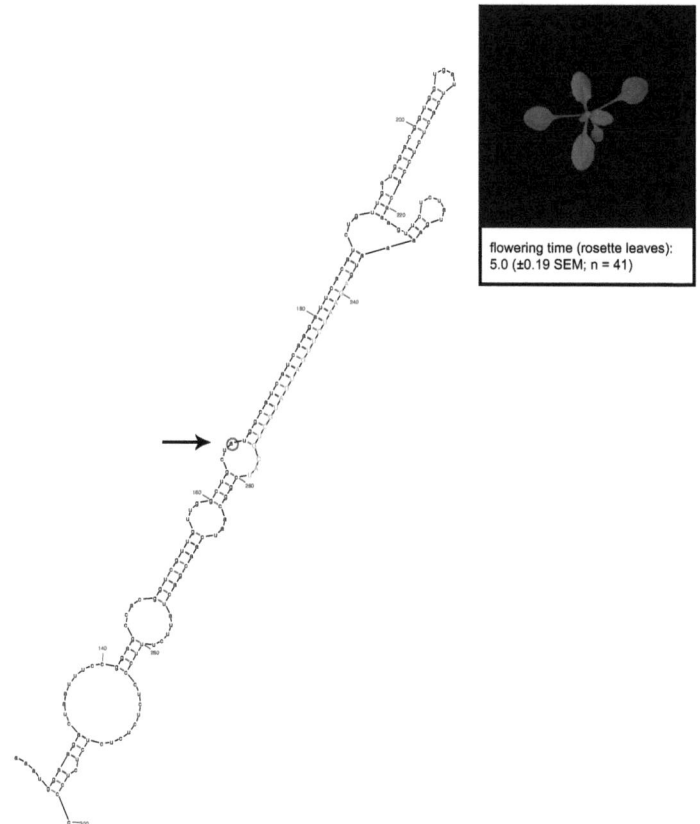

flowering time (rosette leaves): 5.0 (±0.19 SEM; n = 41)

Supplementary Figure 3v
mutant 17

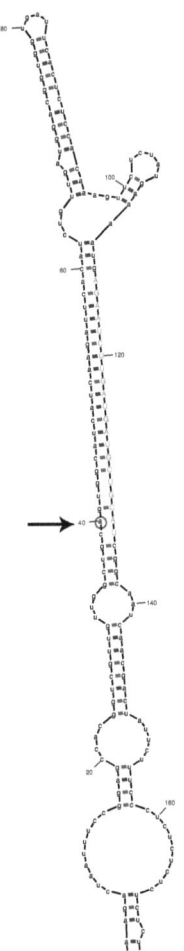

flowering time (rosette leaves):
4.6 (±0.19 SEM; n = 39)

Supplementary Material

Supplementary Figure 3w — mutant 18

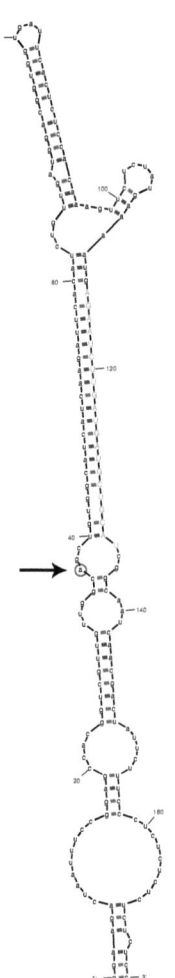

flowering time (rosette leaves):
2.1 (±0.05 SEM; n = 20)

Supplementary Material

Supplementary Figure 3x   mutant 19

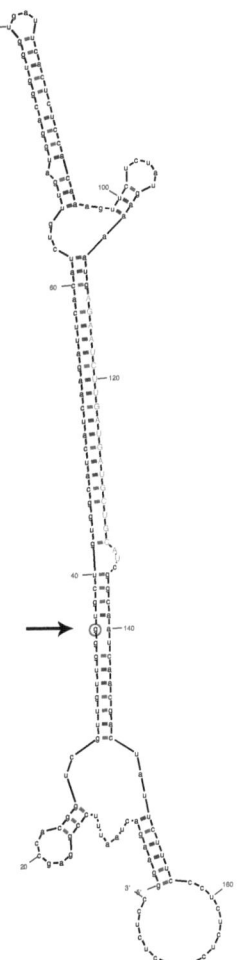

flowering time (rosette leaves):
7.2 (±0.27 SEM; n = 61)

Supplementary Material

Supplementary Figure 3y   **mutant 20**

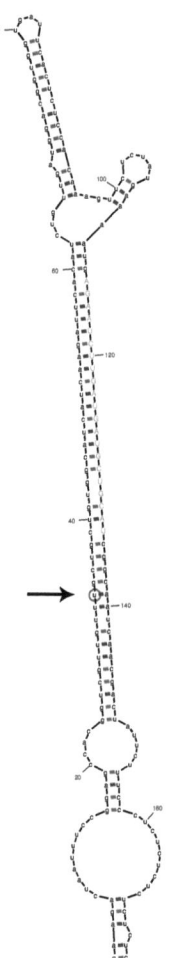
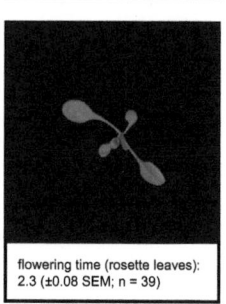

flowering time (rosette leaves):
2.3 (±0.08 SEM; n = 39)

Supplementary Figure 3z   mutant 21

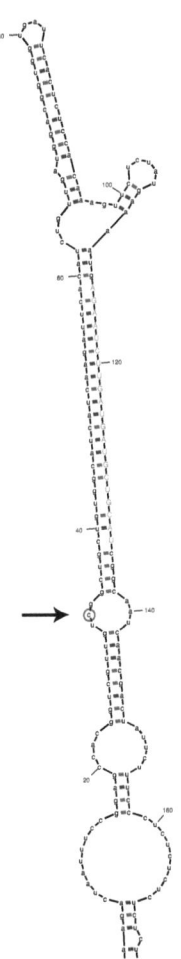

flowering time (rosette leaves):
2.1 (±0.05 SEM; n = 39)

Supplementary Material

Supplementary Figure 3aa   **mutant 22**

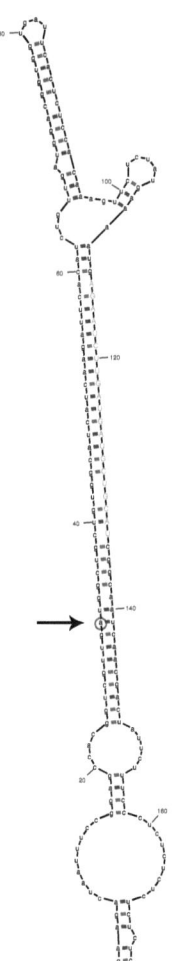

flowering time (rosette leaves):
2.1 (±0.05 SEM; n = 42)

Supplementary Figure 3ab
mutant 23

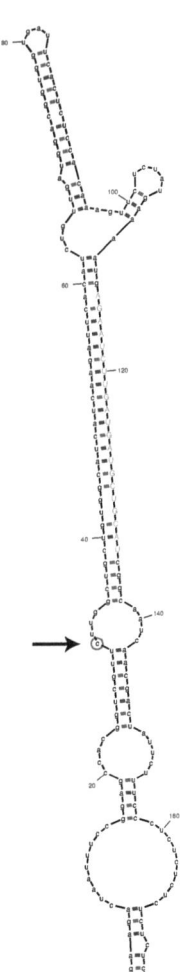

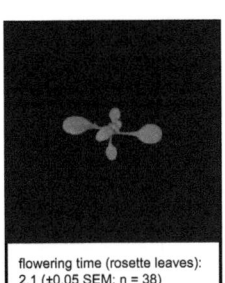

flowering time (rosette leaves):
2.1 (±0.05 SEM; n = 38)

Supplementary Figure 3ac   **mutant 24**

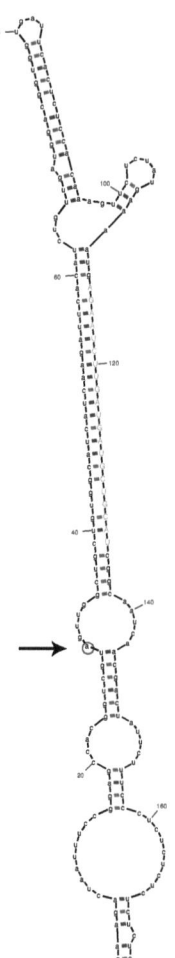
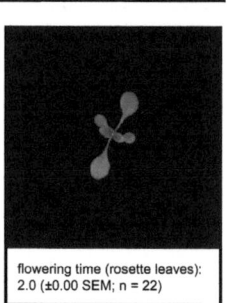

flowering time (rosette leaves):
2.0 (±0.00 SEM; n = 22)

Supplementary Material

Supplementary Figure 3ad   **mutant 25**

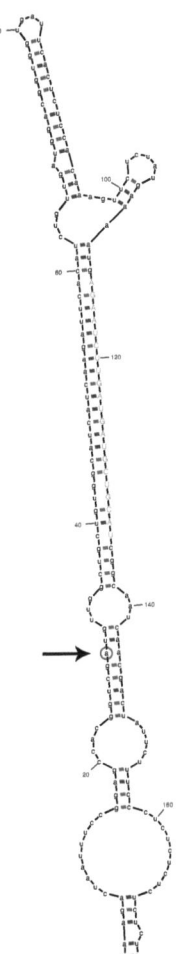
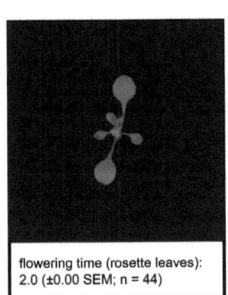

flowering time (rosette leaves):
2.0 (±0.00 SEM; n = 44)

Supplementary Material

Supplementary Figure 3ae   **mutant 26**

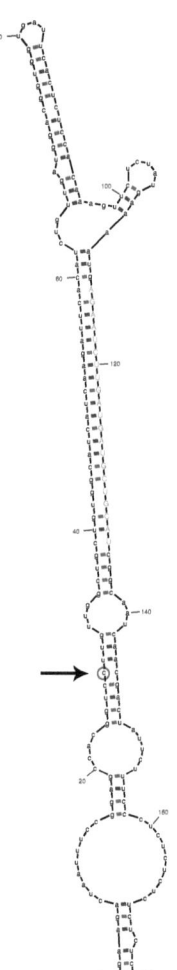
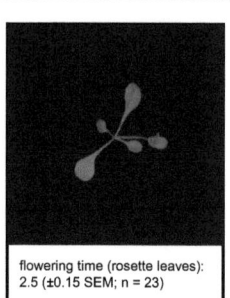

flowering time (rosette leaves):
2.5 (±0.15 SEM; n = 23)

Supplementary Figure 3af

mutant 27

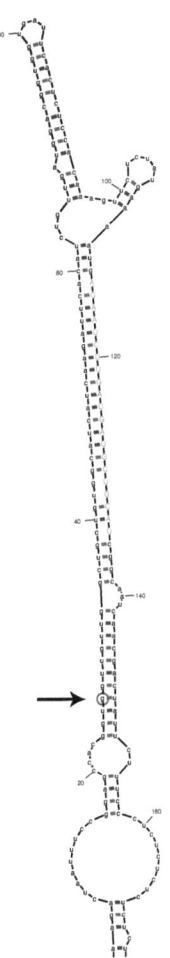

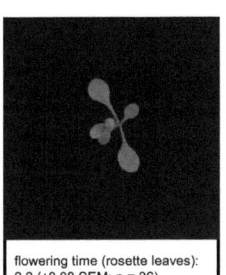

flowering time (rosette leaves):
2.2 (±0.08 SEM; n = 36)

Supplementary Figure 3ag   **mutant 28**

flowering time (rosette leaves): 4.1 (±0.31 SEM; n = 37)

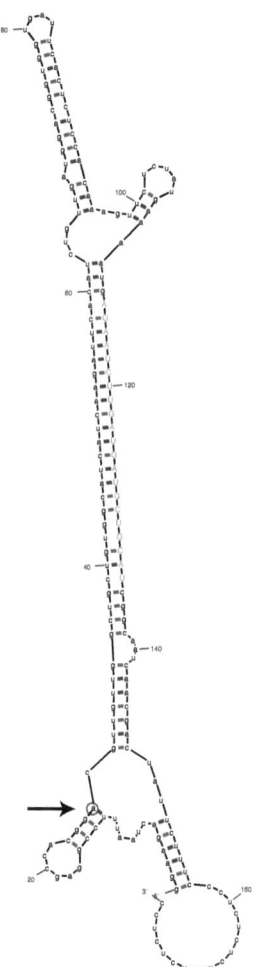

Supplementary Figure 3ah   mutant 29

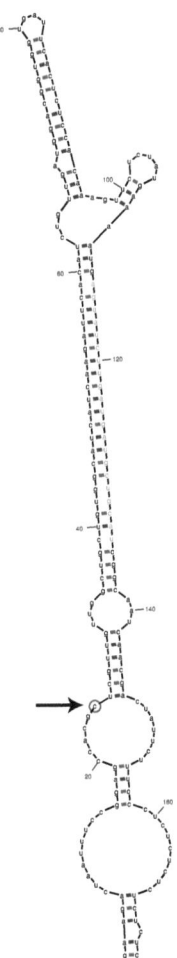

flowering time (rosette leaves):
10.2 (±0.26 SEM; n = 31)

Supplementary Figure 3ai mutant 30

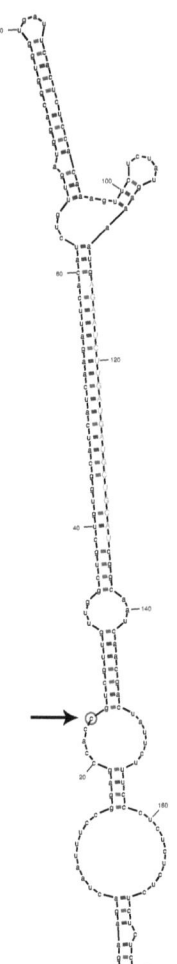
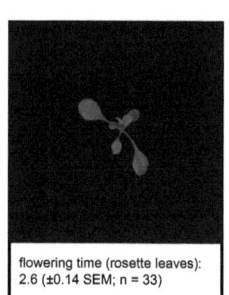

flowering time (rosette leaves):
2.6 (±0.14 SEM; n = 33)

Supplementary Material

Supplementary Figure 3aj   **mutant 31**

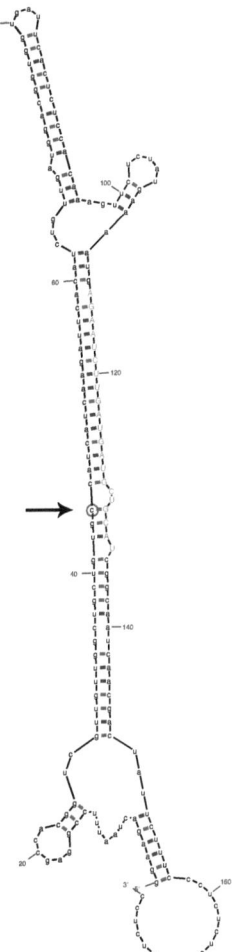

flowering time (rosette leaves):
6.4 (±0.29 SEM; n = 36)

Supplementary Material

Supplementary Figure 3ak   **mutant 32**

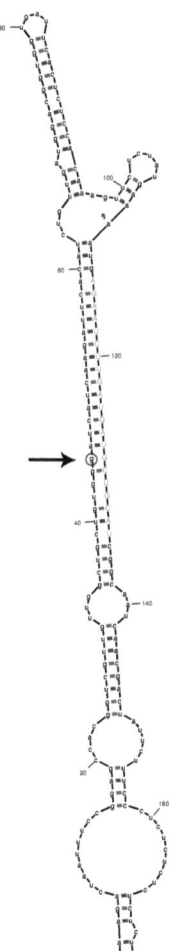

flowering time (rosette leaves):
3.9 (±0.27 SEM; n = 35)

Supplementary Figure 3al   mutant 33

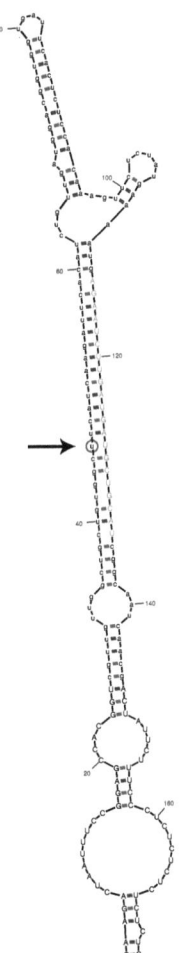
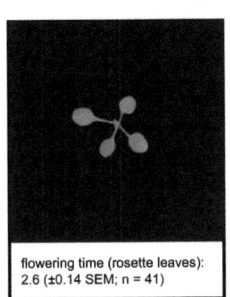

flowering time (rosette leaves):
2.6 (±0.14 SEM; n = 41)

Supplementary Material

Supplementary Figure 3am   **mutant 34**

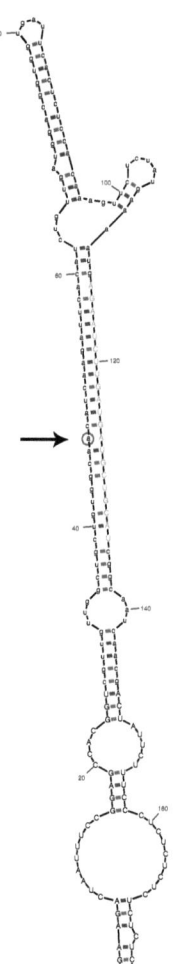

flowering time (rosette leaves):
2.4 (±0.11 SEM; n = 35)

Supplementary Figure 3an

mutant 35

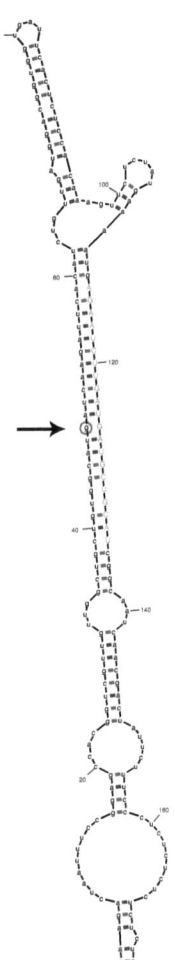
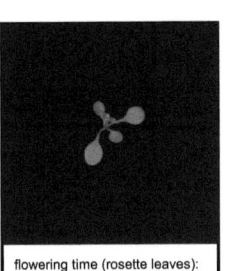

flowering time (rosette leaves):
2.6 (±0.15 SEM; n = 42)

Supplementary Figure 3ao   mutant 36

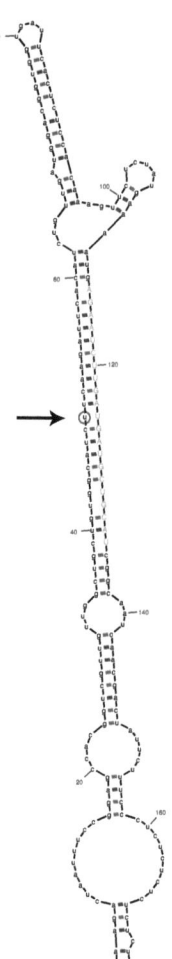

flowering time (rosette leaves):
2.0 (±0.03 SEM; n = 50)

Supplementary Figure 3ap · mutant 38

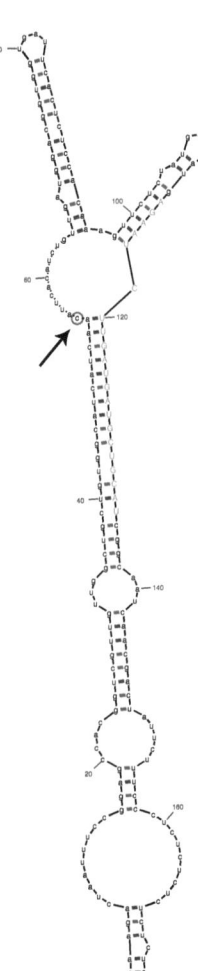
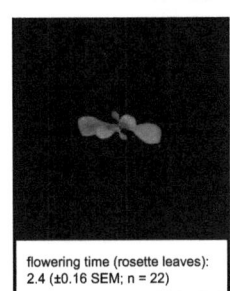

flowering time (rosette leaves):
2.4 (±0.16 SEM; n = 22)

Supplementary Material

Supplementary Figure 3aq   **mutant 39**

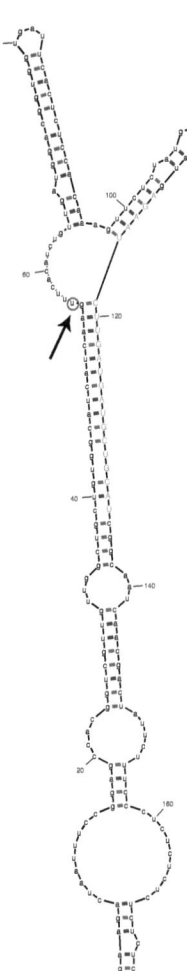

flowering time (rosette leaves):
4.0 (±0.49 SEM; n = 47)

Supplementary Figure 3ar   mutant 40

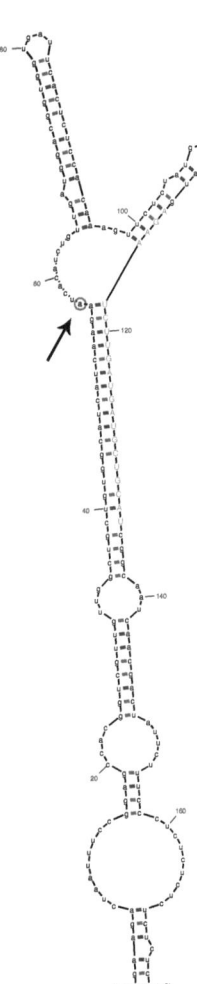
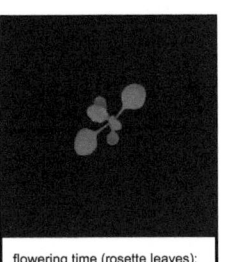

flowering time (rosette leaves):
2.6 (±0.15 SEM; n = 42)

Supplementary Material

Supplementary Figure 3as — mutant 41

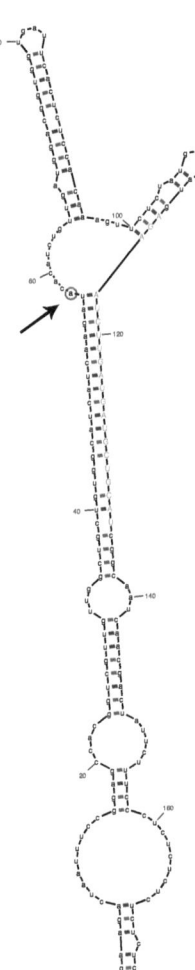

flowering time (rosette leaves):
8.7 (±0.28 SEM; n = 47)

Supplementary Figure 3at   mutant 42

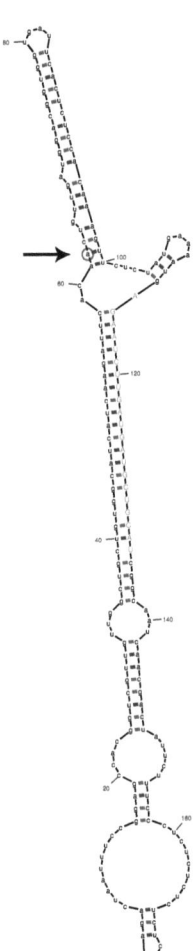
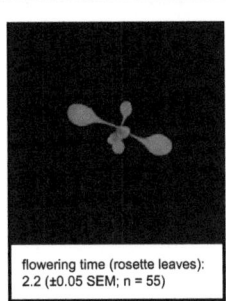

flowering time (rosette leaves):
2.2 (±0.05 SEM; n = 55)

Supplementary Material

Supplementary Figure 3au **mutant 43**

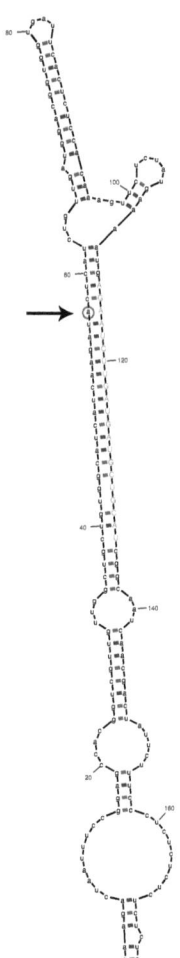

flowering time (rosette leaves):
4.9 (±0.30 SEM; n = 33)

Supplementary Figure 3av   mutant 48

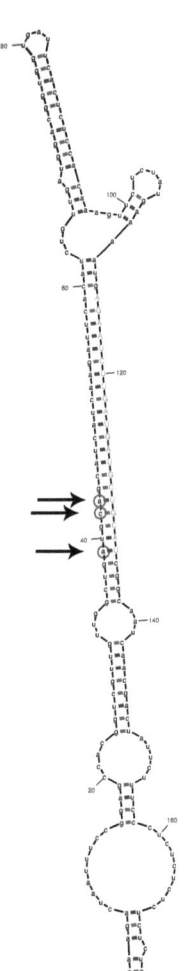

flowering time (rosette leaves):
2.1 (±0.06 SEM; n = 18)

Supplementary Figure 05aw — mutant 49

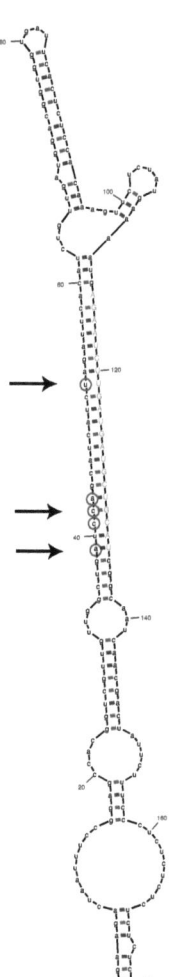

flowering time (rosette leaves):
3.6 (±0.19 SEM; n = 22)

Supplementary Figure 3ax   mutant 50

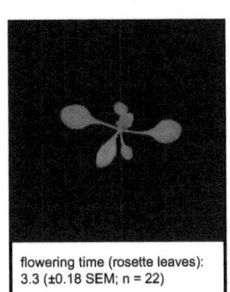

flowering time (rosette leaves):
3.3 (±0.18 SEM; n = 22)

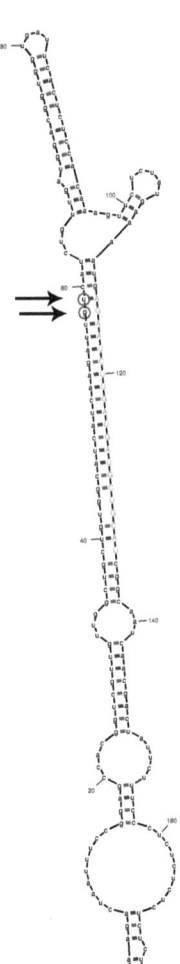

Supplementary Figure 3ay

mutant 52

flowering time (rosette leaves):
2.0 (±0.00 SEM; n = 22)

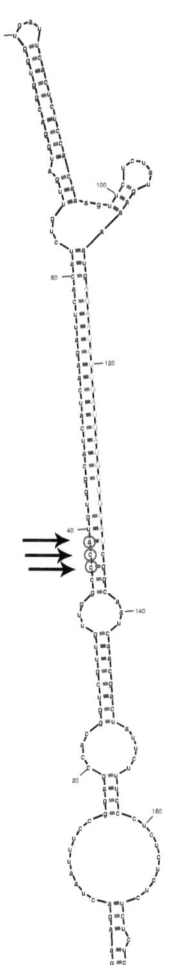

Supplementary Figure 3az   **mutant 53**

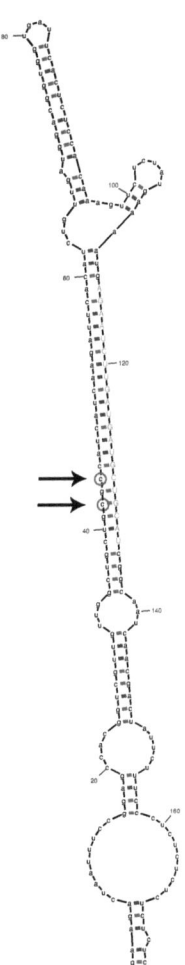
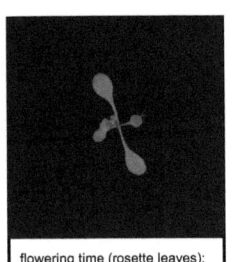

flowering time (rosette leaves):
2.2 (±0.13 SEM; n = 13)

Supplementary Material

Supplementary Figure 3ba **mutant 54**

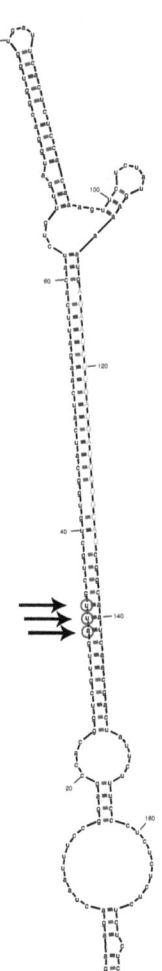
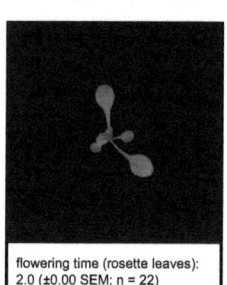

flowering time (rosette leaves):
2.0 (±0.00 SEM; n = 22)

Supplementary Material

Supplementary Figure 3bb   **mutant 55**

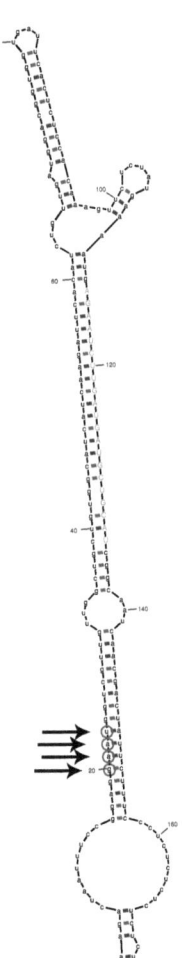

flowering time (rosette leaves):
10.2 (±0.28 SEM; n = 34)

Supplementary Figure 3bc   mutant 57

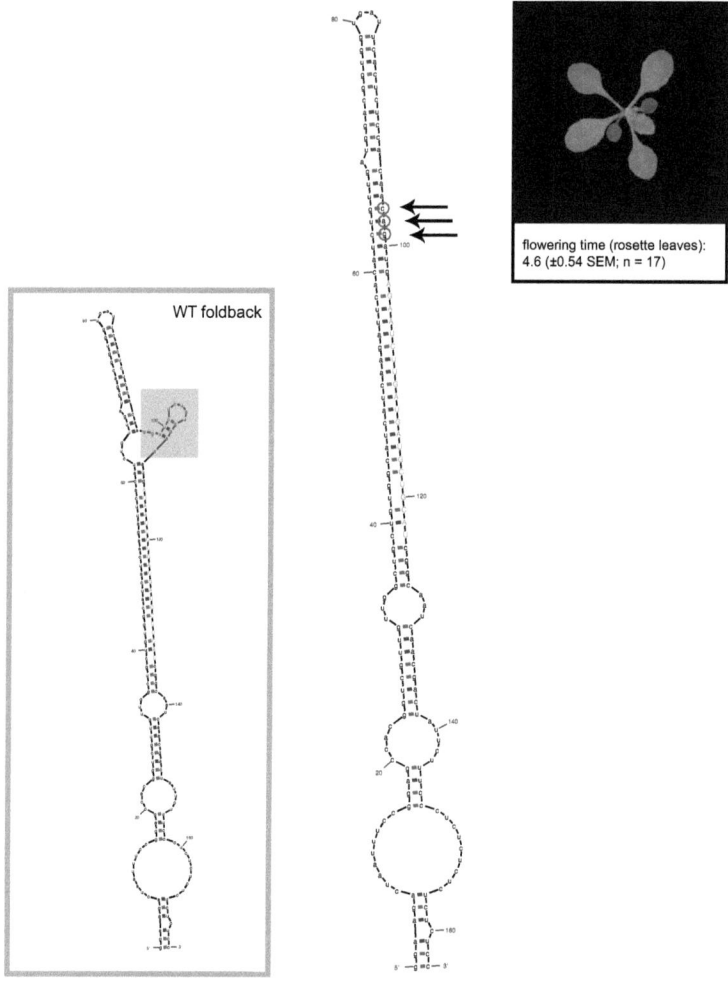

flowering time (rosette leaves):
4.6 (±0.54 SEM; n = 17)

WT foldback

Supplementary Figure 3bd **mutant 60**

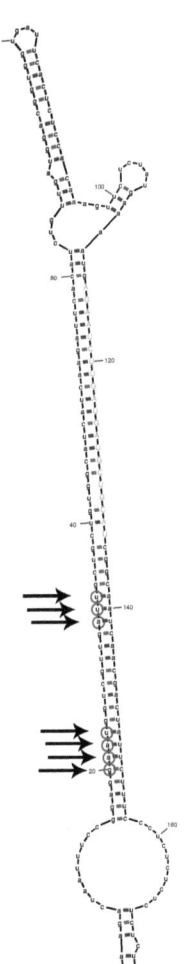

flowering time (rosette leaves):
9.0 (±0.33 SEM; n = 13)

Supplementary Figure 3be — mutant 61

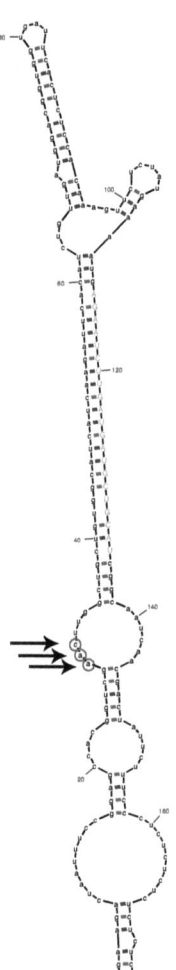

flowering time (rosette leaves):
3.6 (±0.27 SEM; n = 13)

Supplementary Figure 3bf   mutant 62

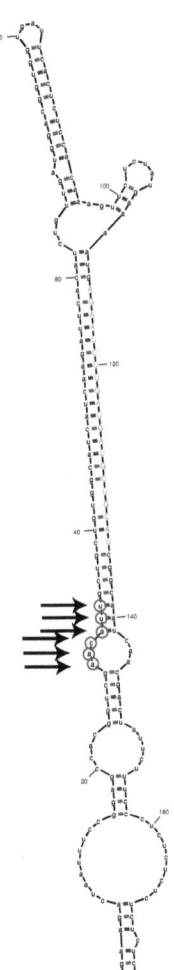

flowering time (rosette leaves):
2.7 (±0.26 SEM; n = 19)

Supplementary Figure 3bg   mutant 63

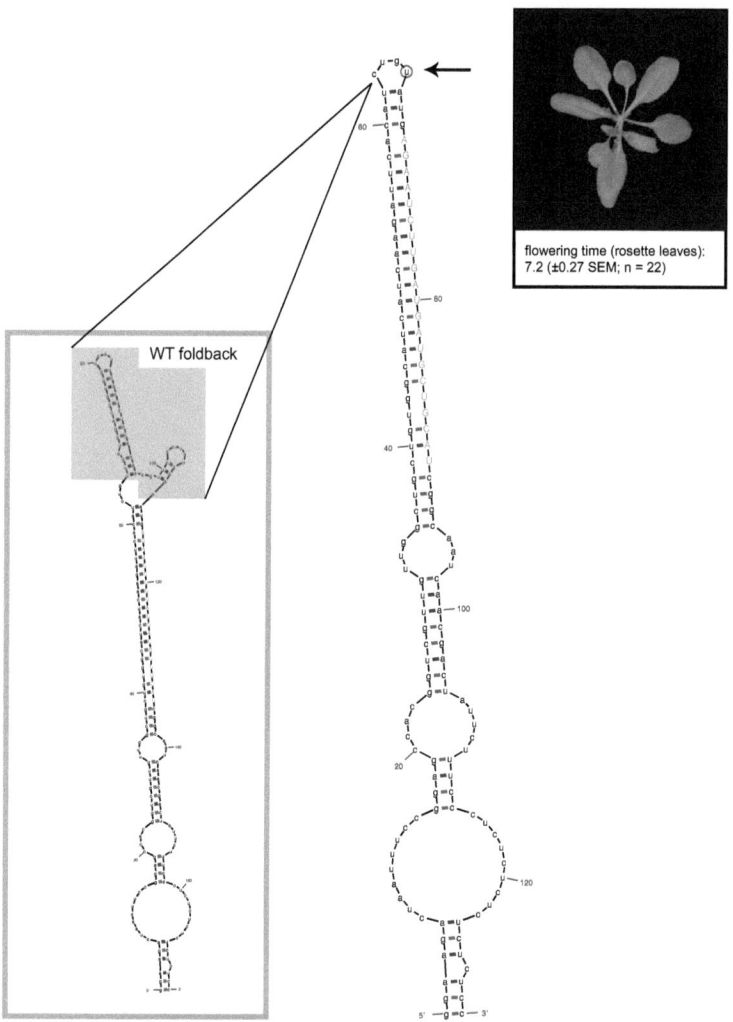

WT foldback

flowering time (rosette leaves): 7.2 (±0.27 SEM; n = 22)

Supplementary Figure 4

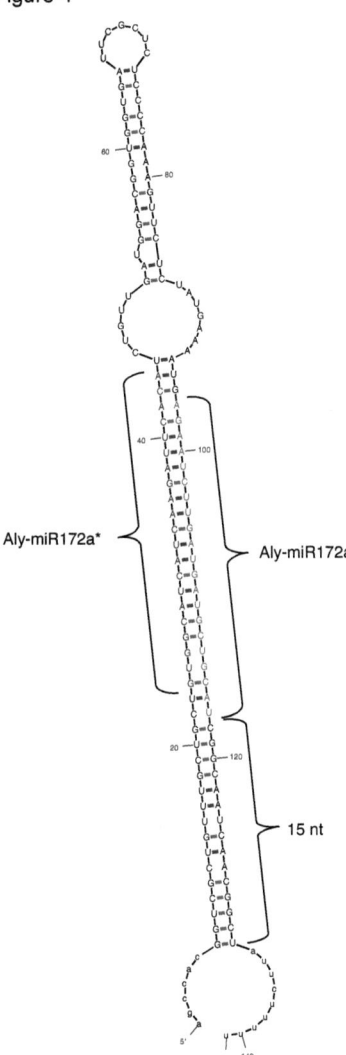

I want morebooks!

Buy your books fast and straightforward online - at one of world's fastest growing online book stores! Environmentally sound due to Print-on-Demand technologies.

Buy your books online at
www.morebooks.shop

Kaufen Sie Ihre Bücher schnell und unkompliziert online – auf einer der am schnellsten wachsenden Buchhandelsplattformen weltweit! Dank Print-On-Demand umwelt- und ressourcenschonend produziert.

Bücher schneller online kaufen
www.morebooks.shop

KS OmniScriptum Publishing
Brivibas gatve 197
LV-1039 Riga, Latvia
Telefax: +371 686 204 55

info@omniscriptum.com
www.omniscriptum.com

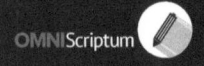

Printed by Books on Demand GmbH, Norderstedt / Germany